THE POLITICAL ECONOMY OF THE OIL IMPORT QUOTA

Yoram Barzel
Christopher D. Hall

Pre-publication review
by Paul H. Cootner, Profess
Stanford Graduate School o

Businessmen and economists seldom see eye to eye. The economist usually views his analytical methods as being eminently practical and useful for understanding the behavior of businessmen and firms. But businessmen usually view the work of the economists as irrelevant academic exercises, which bear no relation to the reality of business decision making. What accounts for conflicting perceptions?

One reason is that in the economist's model of the competitive world, the businessman is viewed as playing a purely passive role, responding to signals from the market but never initiating anything. Businessmen, on the other hand, tend to see themselves as playing an active and crucial role in shaping the world in which they operate. These conflicting perceptions endure because the economist regards the businessman's self-portrait as an illusion, the product of wishful thinking. Can these disparate viewpoints ever be reconciled? Is there a broader model or theory which can integrate these seemingly irreconcilable differences?

The work of Barzel and Hall is an important step toward that goal. They demonstrate one flaw in the standard economic model and in doing so give us a revealing insight into the theory of regulation and the nature of entrepreneurial behavior. In the late 1950s, the U.S. began to administer an oil import quota system—at first "voluntary" and later made compulsory. In the standard view, these quotas are precise, immutable facts of life to the firms affected, and the economist then proceeds to analyze how the firm will react to the intent of the legislators and bureaucrats who administer them. To Barzel and Hall's businessmen, however, these quotas are necessarily imprecise rules, cumbersome to enforce, which stand in the way of transactions that are mutually beneficial to the parties engaged in them. These entrepreneurs feel no compulsion to comform to the nebulous "intentions" of the parties proposing the rules. They have no legal obligation to do more than respond to the letter of the law, and they have positive economic incentives to spend resources to change the rules and to evade its consequences.

In short, the actors on the real stage *do* shape the world in which they operate, and their power to shape it is substantial. "Barrels of crude oil" become as close as possible to barrels of refined products. The benefits of this huge, governmentally-operated cartel are bought and sold through "tickets" for imported oil, and through mergers of inland and coastal refineries. Tank trucks make ingenious, if costly, U-turns across the Mexican border, petrochemical producers erect "foreign" plants in free-trade zones, senators and congressmen are mobilized to press for special exemptions. Regulations are changed again and again and finally disappear altogether.

Taken by itself, this book is an ingenious analysis of the real world of the domestic petroleum industry under the oil import quotas, with all the interplay of governmental edict and entrepreneurial response. But it is more than that. In analyses such as this one lies the answer to the continuing dramas of tax reform and tax loopholes, of law enforcement and "victimless" crimes, of advertising in competitive industries, etc. Barzel and Hall lay the foundations for an analysis of the real world in which people not only produce economic goods but shape economic (and political) institutions as well.

The Authors

Yoram Barzel is Professor of Economics at the University of Washington. He did graduate work in economics at Hebrew University (M.A., 1956) and at the University of Chicago (Ph.D., 1961). Christopher D. Hall is a consulting economist with an M.A. in economics from the University of Washington. Barzel and Hall have contributed to numerous periodicals, including the *Journal of Political Economy, American Economic Review, Quarterly Journal of Economics,* and the *Review of Economics and Statistics. The Political Economy of the Oil Import Quota* is their first book.

THE POLITICAL ECONOMY OF THE OIL IMPORT QUOTA

CE

Yoram Barzel
Christopher D. Hall

THE POLITICAL ECONOMY OF THE OIL IMPORT QUOTA

HOOVER INSTITUTION PRESS
Stanford University Stanford, California

The Hoover Institution on War, Revolution and Peace, founded at Stanford University in 1919 by the late President Herbert Hoover, is an interdisciplinary research center for advanced study on domestic and international affairs in the twentieth century. The views expressed in its publications are entirely those of the authors and do not necessarily reflect the views of the staff, officers, or Board of Overseers of the Hoover Institution.

Hoover Institution Publication 172

International Standard Book Number: 0-8179-6721-4
Library of Congress Catalog Card Number: 76-41087
Printed in the United States of America

Contents

ILLUSTRATIONS vii

TABLES ix

PREFACE xi

ACKNOWLEDGMENTS xiii

INTRODUCTION 1

1. THE QUOTA'S SETTING: A BRIEF HISTORY 7
   - How American Is American? 8
   - Overland Imports 10
   - The Special Treatment of District V 14

2. A VOLUNTARY (?) QUOTA 16

3. THEORETICAL FRAMEWORK 22

4. THE POLICING OF THE QUOTA 31
   - The Petrochemical Challenge 43
   - Two Ad Hoc Examples 49

5. THE ALLOCATION OF CRUDE OIL: OIL SPOILS 51
   - Refining and Historical Allocations 51
   - The Effect of the Refiners' Allocation 53

6. THE EFFECT OF THE QUOTA ON THE STRUCTURE OF OIL PRICES 65

7. SUMMARY AND CONCLUSIONS 72

APPENDIXES 75

A-1: Import Allocation Formulae 76

A-2: The Formal Definitions 81

B: Some Posted Prices 88

BIBLIOGRAPHY 93

INDEX 95

Illustrations

FIGURE 1. Canadian Imports from and Exports to the United States 11

FIGURE 2. Simple Model of an Import Quota 23

FIGURE 3. Value of a Ticket to Import Oil under the Quota 28

FIGURE 4. Equilibrium Price of Ticket under Competition 57

FIGURE 5. Midwest Refiners with Access to Canadian Crude 64

FIGURE 6. Ratios of Light Products to Fuel Oil 67

FIGURE 7. Import of Unfinished Oils as Fraction of Crude 69

Tables

TABLE 1. Effect of Quota on Relative Price 39

TABLE 2. Refining Capacity by Refiner Size (Districts I-V) 54

TABLE 3. Refining Capacity by Plant Size (Districts I-IV) 55

TABLE 4. Refining Capacity by Refiner Size (District V) 56

TABLE 5. Refining Capacity by Plant Size (District V) 56

TABLE 6. Share of Regional vs. Interregional Firms 62

TABLE 7. Inland Refineries Owned by Refiners with Capacity on the East Coast 63

TABLE 8. Crude, Unfinished, and Products Imported into the United States 68

Domestic Studies Publications

In 1971, the Overseers of the Hoover Institution approved a major expansion in research on domestic policy issues. The objective of this action was to bring scholarly effort to bear on important problems of public policy. The output from this research has been reflected in numerous articles and monographs. This book inaugurates a new series of publications which stem from the growing research sponsored by the Domestic Studies Program.

Thomas G. Moore, Director
Domestic Studies Program

Preface

Until a few years ago, the economic theory of regulation had a remarkably simple, and naive, structure. A law or a regulatory decree was issued after a political discussion, and whether it represented the triumph of special or general interests, it was an exogeneous phenomenon from the viewpoint of economics. There was, to be sure, some trifling investment of time and intelligence in the debate preceding the regulation, but this trifling investment bore no relationship to the magnitude or complexity of the policy which was promulgated.

Then, once the regulatory policy was established, the economist analyzed its impact upon output, price, quality, or any other variables the policy was intended to influence. Individuals and firms adapted to the regulations as parameters of their future conduct, exactly as a competitive wheat farmer adapted to the price of wheat. The price ceiling or the import quota or the quality minimum was adapted to by the profit-maximizing firm, and that was that.

This simple picture of regulation has been changing rapidly in recent years, and for the better. The regulatory policies are no longer viewed by economists as data—as ordinances issued by some modern Moses. These policies are the product of sustained, purposive campaigns by importantly affected parties: a tariff increase or decrease is the attained end of a concerted effort by the protected industry and its allies, or by the affected customer industries and their allies.

In a careful study of the largest public cartel in world history, Barzel and Hall carry this new branch of analysis to new depths of understanding. They do not undertake to analyze the appearance of the oil import quota system in the late 1950s—that is a task still to be performed. Barzel and Hall begin at this point in time and examine the variety of changes in the policy that occurred before its demise in 1973. For example, there were frequent changes in quotas, major redirections of policy toward petro-chemical importers, residual fuel importers, and other groups, and they illuminate these changes by examining the costs and benefits to various parties of these policy changes as the underlying market conditions developed.

It is in the second part of the theory of regulation, however, that this study makes its largest contribution. A regulatory policy such as the oil import quota system deals with products (crude petroleum), firms (refiners), areas (coastal firms), and quotas (barrels per day) that have customarily been accepted as the data of economic analysis. They show, in rich and illuminating variety, that these categories are the beginning, not the end, of economic analysis. Products, firms, areas, and quotas are no doubt ambiguous when left to themselves, but the ambiguity is increased as the result of profit-seeking activities of the various participants in the market.

Thus they show how the quota system led to increases in high quality oil imports, and these imports in turn led to important changes in relative prices of oil products. They show how the market in "tickets" to import oil led to extensive mergers of inland and coastal refineries. No one who reads Barzel and Hall's fertile and ingenious exploration of the dimensions of each regulatory variable will ever again look upon the economic analysis of a regulatory policy in quite the same way.

George Stigler
May 10, 1976

Acknowledgments

We wish to thank Wilson Mixon for his assistance in the collection and compilation of data and documents. Steve Sobotka provided us with numerous insightful comments on an early draft.

Thanks are due to the Bureau of International Labor Affairs, U.S. Department of Labor for financial support. Views expressed and conclusions reached do not necessarily represent those of the Department of Labor.

Introduction

Though the oil import quota constituted a transient episode in American economic history, extending only from the late fifties into the early seventies, the dollar magnitudes involved were large. Roughly, crude oil production in the United States stood at nine million barrels a day, with imports adding another million barrels a day.[1] The East Coast price of domestic oil was \$3.00–\$3.50 a barrel, or more than one dollar higher than the delivered cost of imported oil. These annual magnitudes of several billion dollars did not go unnoticed by those in or close to the industry.

The economics literature is meager on the oil quota in particular and on quotas in general, and we might speculate on the reason for this paucity.[2] The textbook analysis of a quota sets a framework greatly resembling that for a tariff and, in fact, offers a proof that under competition, except for distributional considerations, the two can be made identical.[3] In turn, the analysis of the effects of tariffs (and of other taxes) seems quite simple to the textbook writer. Therefore he sees no need for much specific discussion of quotas.

Most of the economic literature on the oil quota is devoted not to conceptual problems—these are usually dealt with in a couple of paragraphs—but rather to estimations of quantities, such as the cost of the quota to consumers.[4] For calculations of this sort, estimates of domestic and foreign supply schedules and of the home demand schedule are almost the only components.

It may come as a surprise, at least to economists following this line of inquiry, that the oil industry took keen interest in the subject of an import

1. *Minerals Yearbook,* U.S. Bureau of Mines (Washington, D.C.: Government Printing Office), relevant years.

2. Consider, for instance, the survey by R. M. Stern, "Tariffs and Other Measures of Trade Control: A Survey of Recent Developments." In a total of 136 references, the word *quota* appears only once—in Number 15, by Bhagwati. (See next footnote.)

3. See J. Bhagwati, "On the Equivalence of Tariffs and Quotas," in his *Trade, Tariffs, and Growth.*

4. A major example is *An Analysis of the U.S. Oil Import Quota,* by J. C. Borrows and T. A. Domencich.

quota. The *Oil and Gas Journal,* a major oil periodical and a champion of oil interests, found it worthwhile to editorialize constantly and to devote space both to numerous lengthy articles and to a huge number of news items dealing with the quota. What was there to report and to comment upon?

It is our contention that economists, in an attempt to simplify and abstract, have overlooked essential features of the problem; thus their results have been both incomplete and biased. As an illustration, consider the commodity known as crude oil—the centerpiece of the oil import quota. A barrel (of forty-two U.S. gallons) is its usual measure, and the quota is stated in that unit. Crude oil is accepted implicitly as a homogeneous commodity for which diagrams can be drawn and estimates can be derived in terms of the number of barrels and the price per barrel. At best, some lip service is given to the real-world possibility of heterogeneity.

But is crude oil actually homogeneous enough to be treated as a single commodity? The oil extracted even from a single field is not entirely uniform, and its characteristics can be substantially modified by varying the rate of output and by the way associated gas is treated, as well as by employing various secondary recovery methods. Across fields, the differences in the characteristics and values of crude oil are still greater.[5] It may, then, matter a great deal whether the quota is specific as to the quality of its commodity or whether, as is actually the case, its terms refer simply to barrels, presumably of any crude oil.

This still is an oversimplification. When does oil cease to be crude and commence to become refined? Simple and low-cost methods allow crude oil to be partly processed at the field.[6] Such techniques are not normally classified as refining, yet they change the value of the oil considerably. Indeed, at any time, each of many oil fields in the United States offers oil of varying qualities. Within a single field, the highest per-barrel price of crude may often exceed that of the lowest by some 30 percent and sometimes by much more.[7] One may search in vain for such a major quantitative and qualitative distinction in most economic studies.

Though this problem has been overlooked by economists, some have felt impelled to identify studiously the unit of measurement. A recently established futures market for oils found this concern to be more than pedantry when the market executives attempted to establish what was to be traded on their exchange. The *Petroleum Economist* provides this description:

5. In Shakespeare's case and in Gertrude Stein's, since "roses" are neither subsidized nor taxed, the author(s) committed no conceptual error.

6. For instance, certain fractions of the crude may be separated and returned to the underground pool, thus changing the quality of both the marketed oil and that remaining in the pool.

7. See Appendix B for a sample of price variations within and across fields.

> The technical specification of the basic contract is 5000 barrels of 42 U.S. gallons each with API gravity of 34° and a sulphur content of 1.7 per cent, bottom sediment and water maximum 1 per cent.
>
> Premiums and discounts are established for other crudes deliverable, provided they fall within a gravity range of API 27°–API 45° and a sulphur content of minimum 0.1 per cent and maximum 3 per cent. All international crudes meeting these standards will qualify for trading, as also the 40 per cent of U.S. domestic production specified by the Administration as "new oil" (oil from wells brought in since 1972 or from an old well over and above the well's 1972 production). . . .
>
> *Variety of Crudes.* A peculiar difficulty in setting up the market has been the multiplicity of qualities of crudes, and an ingenious solution has been arrived at. The Cotton Exchange has arranged a committee, including representatives from different sections of the oil industry, who will make recommendations by telex to the Exchange in the first week of the month preceding a delivery month as to their views on appropriate premiums and discounts from the base contract for oils falling within the range described above. The Exchange will average differences of opinion, and publish the results on the 10th day prior to delivery date, the general idea of course being to allow for variations in differentials reflecting changes of season and customer demand.[8]

Additional testimony on the heterogeneity of barrels can be found in the records of oil exchanges, or in the so-called ticket swaps, where *tickets* refer to the right granted by the quota to import one barrel and *swaps* describe an officially sanctioned method of cashing in one's allocation of import rights. We discuss this process in greater detail later, but for the moment it suffices to note two prominent features of these data: (1) The number of barrels only partly describes the transaction, as a quality specification also accompanies each quantity variable. (2) The value of a given quantity is directly related to its quality (100 percent variation in price per barrel is not uncommon).

Since the quota is actually stated in terms of so many barrels of (any) crude oil, the economist who ignores variability implicitly assumes that the imported crude is to be of average quality. We predict, however, that the quota will induce importation of higher quality oil, as will be discussed in greater detail subsequently. How does the change in quality relate to conventional estimates of resource misallocation?

The demand elasticity for oil may be estimated from data generated independently of the quota. In the absence of known incentives to modify quality, it seems reasonable to assume that the data are for fairly uniform quality. An estimate of resource misallocation due to the quota, if based on such elasticity, will be biased upward since it fails to take into account the market's incentive for quality adjustment. On the other hand, if the data used

8. *Petroleum Economist,* October 1974, p. 365.

to obtain the elasticity estimate are from a comparison of the prequota with the postquota periods for observed numbers of barrels and observed (average) prices, the bias is reversed. In terms of the resource cost, what is overlooked in the latter case is that the quota has induced a costly change in resource use, since the new quality is not optimal.

The composition of crude oil is only one of many dimensions subject to adjustment under the import quota program. No formulation of a quota can avoid all ambiguities, and market participants will spend resources to take advantage of the emerging opportunities in one way or another. These costs cannot be captured in the conventional formulation of demand and supply relations.

Neither the formal statement of the quota nor its actual administration is by divine decree. The so-called mandatory quota imposed by presidential proclamation in March 1959 was by authority of the Emergency Preparedness Act and presumably for defense purposes. Even if one accepts that justification at face value, it is clear that the quota had profound implications for oil producers and consumers, who consequently were willing to spend real resources in an attempt to shape its provisions and implementation in line with their interests. Activities of this kind undoubtedly began well ahead of the quota itself, as various market participants maneuvered to promote (or to protect) their causes.[9] Resources are consumed in this type of project, and the extent of their use is governed by the principle that, at each of the relevant margins, one dollar in cost will yield one dollar in benefits.

The initial date of the quota proclamation and its specific nature imply a certain alignment of interacting economic and political forces. The emergence of the quota signals that the spending of resources to obtain a favorable political decision does indeed yield a return. We should expect, then, that changing economic and political configurations over time would also produce changes in the quota specifications, with each new shift implying both that resources have been expended to effect it and that new resource allocations will be made. As we shall see, such changes did take place—at times with dizzying speed.

The fundamental premise of this paper is that the oil quota provides numerous margins in addition to quantity along which market participants can and will adjust to the point where further realignment no longer yields additional net benefits. The total of the individual adjustments has, on the market level, one important and simple property: given the quota constraints (i.e., the quota as actually implemented) and the costs of transactions, adjustments by the industry will be such that the dollar value of resource misallocation induced by the quota is smallest. Since the quota establishes

9. Actually several voluntary import quotas were in effect in the late fifties; each of those, we suspect, induced a similar flurry of action.

only an incomplete set of property rights, the unspecified components will be claimed by way of competition, and the "efficient" market solution will be to minimize the resource cost of that process.

The adjustment to the quota, as we see it, is much more involved than that usually portrayed, and to pinpoint its effects requires a large amount of detailed and accurate information. The available data in the oil industry are quite unsatisfactory, possibly not entirely by accident. For that reason, much of our evidence is qualitiative rather than quantitative; while we can often point to the direction of the adjustment, and sometimes may be able to indicate its importance, we will seldom be in a position to assess its magnitude accurately.

Both the industry and the government have numerous incentives to distort their measurements.[10] The industry obviously may benefit from a bias in the accounting (a subject that will be addressed later in more detail). Governmental statistics based on the industry-produced data will reflect the same bias. Further, accurate data are expensive to collect, and their derivation requires that the producer have an incentive to measure the same quantities as the investigators.

Posted prices are a prime example of data that are difficult to interpret, since they may bear little resemblance to actual exchange prices at any given time.[11] Some of these postings, such as for Saudi Arabia Light, remained absolutely constant for several years at a time, while tanker rates rose and fell to levels that alone would have forced market prices to follow suit. Since such prices are asserted for the purposes of computing taxes and royalties owed to the various governments, changes in them must be negotiated between the producer and the controlling government and involve substantial costs.

In our research we have used, in addition to the standard publications, two basic sources in an effort to reduce the "noise" occasioned both by the costs of accuracy and its apparently low payoff, and by the systematic bias due to taxes and subsidies. We have relied heavily on industry journals and publications for the spotting of adjustment margins and, to a lesser extent, for indirect measurement of how costly these adjustments were. The accumulated statements of the controlling agencies involved in the quota provide the other source. These include the findings and appeals to the Oil Import Administration (OIA) and the record of judgments of the Oil Import Appeals Board (OIAB).

---

10. The definition of the quantity one desires measured can never be complete. On this front too, then, an individual will choose that variant of the measure that maximizes his net worth. The result may deviate significantly from the measure best suited for testing of hypotheses.

11. Adelman in his *The World Petroleum Market,* spends considerable time constructing various price series for oils, noting that posted prices are merely a method for calculating taxes. Unfortunately for our purposes, his own data do not cover the range of types of oils and of time that we require for testing the hypothesis presented.

Obviously, any sensible analysis of problems associated with the quota requires some knowledge of the physical characteristics of the commodity under consideration as well as of the institutional setup related to it. Similarly, for the derivation of testable implications, one must know the costs and benefits of various quota features to the market participants.

It is evident that our knowledge is spotty and that training in economics alone does not provide the overall expertise needed. Indeed, not even an economist specializing in oil problems for many years could claim mastery of such a complex industry. Our hope here is that our readings of industry reactions to various quota measures will shed light on some relevant issues and reveal certain essential margins. The discovery of such margins is the goal of the bulk of this book.

Where we cannot fully explain some aspects of the formulation of the quota or of industry reactions, but where we nevertheless feel that some important systematic elements can be detected, we will try to offer as coherent a story as possible, with hints of where we feel the answers lie.

1 • The Quota's Setting: A Brief History

The mandatory quota proclaimed by President Eisenhower in March 1959 was preceded by a series of voluntary quotas, each more imperative than the previous one and each only partly successful in slowing down the growth in the import of crude oil. These programs succeeded in restricting some importers but not others, and they failed to control substitutes, particularly petroleum products and unfinished oils.[1] The mandatory quota not only covered these two classes of substitutes but also set their imports at a level drastically lower than in 1958.

The mandatory program can be analyzed along various lines. Separate programs were established for crude oil (also covering unfinished oils) and for petroleum products. The latter area was further divided into light products (mostly gasoline, jet fuel, and distillate) and into heavy or residual fuel oil. Crude oil imports were set at approximately 10 percent of crude consumption and essentially were maintained that way throughout the life of the program. Light product imports were set at a low, and superficially fixed, level. In actuality, however, imports of all but crude oil were allowed to grow quickly.[2]

The country was divided geographically into two main (defense) regions. Districts I–IV, and District V, the first group comprising all of the nation except the Pacific states, Nevada, and Arizona. The two areas operated under differing programs, the regulation of crude oil being more restrictive in the East.

The exporters of the world were also divided into camps: Canada received the most preferential treatment, Mexico was tossed a crumb of the overland exemption, and the remainder of the western hemisphere was in general favored to a degree over the eastern hemisphere, but there were variations through the period.

Qualification for the crude quota was initially based on two alternative criteria: whether the candidate was refining domestic crude, and whether the candidate could show a record of past imports. For petroleum products, the decision was based entirely on historical precedent. Both sets of criteria were modified in the mid-sixties.

---

1. For more detail see Kenneth W. Dam, "Implementation of Import Quotas: The Case of Oil."
2. See Table 8, p. 68.

Management of the quota was relatively even-handed until late in 1965, thus spanning the end of the Eisenhower administration, the whole of Kennedy's, and the first Johnson term. Violent changes erupted during the second Johnson term and, in different directions, through the first Nixon term. The quota was basically terminated in 1973, early in the second Nixon term.

The first major modification, announced on January 1, 1966, gave petrochemical firms a share of the crude oil quota for the first time. A subsequent set of changes accorded special treatment to individual firms, largely those operating in Puerto Rico and the Virgin Islands.

At about the same time, heavy fuel oil went through a series of decontrol steps. The quota on light fuel oil was raised by the mid-1960s. Later, as an incentive to reduce sulphur levels, the import of crude oil was allowed for the first time to exceed that of the established formula. The Oil Imports Appeals Board, which previously had served mostly to turn down requests and to correct inequities, was now handed the task of allocating a substantial share of the quota.

The Nixon administration concentrated much of its import control effort along the Canadian border, where a special Canadian quota was imposed except with respect to District V. This allocation went through a series of gyrations until other forces came to the rescue.

In May 1973, a system of tariffs was established, and imports were decontrolled. At the same time, however, as a final vestige of the quota, all those who had imported oil in the preceding period were granted a tariff exemption that was slated to be slowly phased out.

HOW AMERICAN IS AMERICAN?

The quota restricted oil imports into the United States. That may seem a self-evident designation. But with the quota, as with travel, it soon became evident that Canada and Mexico are not as non-American as is, say, Japan. The special treatment under the quota that was granted to the two neighbors will be discussed in detail presently.

Now, how about Puerto Rico? By certain criteria Puerto Rico could be classified as part of the United States, but not by others. Although this problem was explicitly recognized when the quota was initiated, a few years later the original distinction proved inadequate and special problems caused by that ambiguity contributed largely to the change in resource allocation effected by the quota. The Virgin Islands, and to a minor degree Guam, caused similar problems.

Even within the more narrowly defined geographic limits of the United States, certain ambiguities remained. First is the problem of bond—of imported commodities to be reexported. Large quantities of fuel are used to propel international travel and trade. Should such fuel be subjected to quota limitations? While the oil industry was concerned about this issue, it seems that a relatively liberal policy was adopted from the very beginning: ships and airplanes traveling internationally were allowed to use bonded fuel exempt from quota restrictions.

This relaxed attitude was not entirely surprising. Given both the price difference between local and foreign fuels and the access of international carriers to foreign fuels, particularly for marine vessels, the effect of quota restrictions on this part of the market was bound to be reduced.

Put differently, the demand by international carriers for domestic oil appears relatively elastic, for they can more readily substitute foreign for domestic oil (or, more accurately, oil products). The more elastic the demand, the higher the resource cost per dollar gained by owners of the quota, and the stronger the resistance to the quota contraint. Note also that the cartelization of international carriers may render the resistance more effective.

Such was not the case with respect to free trade zones. These areas, while physically on U.S. territory, are beyond the jurisdiction of U.S. customs. Importers use such areas either for storage until they choose to pass the merchandise through customs, or as a site for processing the imports so that duties will eventually be paid on processed rather than on raw commodities.[3] Since refined oil was subject to basically the same quota restrictions as crude, there was little incentive to refine oil in the free trade zone.

Unlike refined oil products, petrochemicals were not covered by the oil import quota. Rather they were, and still are, a major U.S. export. Until the mid-sixties, the U.S. petrochemical industry had used natural gas as its main petroleum raw material, but in that period crude oil and its derivatives became more economical than gas and suddenly the oil import quota represented a costly obstacle to the industry.

The first attempts to escape its effect were in proposals to construct petrochemical plants in free trade zones. It appears that permission of this kind was being granted routinely in other cases; nor did the processing of oil in a free trade zone violate any law or regulation, including those of the quota. Nevertheless, the requested permits were never granted. Instead, after many months of lobbying, the industry was handed a share of the quota, in a solution that proved unsatisfactory from virtually any point of view as we shall see later.

Such treatment of the free trade zone is particularly illuminating on the score of resource allocation. We are unable to estimate the costs to the petrochemical industry, and subsequently to some refiners, of their efforts to gain permission to operate in free trade zones. The stakes involved, however, were substantial. For instance, Union Carbide, the first to try the ploy,

3. These zones also substitute for bonding, since what is reexported is treated as if it never had entered the country.

requested a permit to import some 10,000–15,000 b/d (barrels per day) to a plant to be constructed in Taft, Louisiana, in an area to be classified as a free trade zone. Throughout the sixties the right to the quota was valued at about $1.25 per barrel, so the arithmetic indicates that the value of the permit was around $5 million a year.

The economics of the situation are somewhat more complex. What was the chance of obtaining a permit? In any given case, the bureaucrat on whom such a decision rested might base his answer on any of a dozen or more considerations, including the choice of site, the size of plant, the source of raw materials, or the urgency of the need. Consequently, we cannot expect that exactly the same operations would have been undertaken had the required permit not been subject to such considerations. More generally—to maximize a firm's present value—on the margin, an increase of one dollar in the anticipated value of the permit would have called for one dollar of expenses. (For the project just mentioned, since no information is available regarding the change in resource use induced by the quota, the foregoing statements are mere assertions. Later on, some supporting data with respect to similar projects will be provided.)

OVERLAND IMPORTS

Canada. The presidential proclamation of March 10, 1959, setting the mandatory oil import quota contained no explicit provisions regarding Canada. However, its first amendment, on April 30, exempted overland shipments from quota restrictions—at the expense of other sources, since the overall level of imports remained unchanged.[4]

The overland exemption was justified on the plausible ground that for security purposes North America should be considered a single unit in terms of oil-producing capacity. While there is no doubt that the quota restriction increased the share of domestic U.S. production in total U.S. consumption, the effect of the overland exemption on the corresponding combined share is less self-evident.

Most Canadian crude oil is produced in Alberta. The main natural market for this oil is the neighboring provinces and the U.S. Midwest and West, with some of the oil nowadays passing by pipeline through the American Midwest to Ontario. Much of the eastern Canadian area, however, can be supplied more cheaply from Venezuela, the Middle East, and Africa. Since the overland exemption changed only the sources of imported oil to the United States, not its total volume, its effect on U.S. production was presumably minor and indirect. Changes in the geographic price pattern would alter

4. That the amendment applied directly and wholly to Canada is evident; there is no Mexican production near the border. Nevertheless, Mexico was indirectly affected to a small degree, as will be discussed farther on.

somewhat the geographic composition of output but total production would hardly be affected. The exemption, then, could not have resulted in a significant change in the combined crude oil output unless it increased the Canadian output.

FIGURE 1

Canadian Imports from and Exports to the United States, 1950-1970.

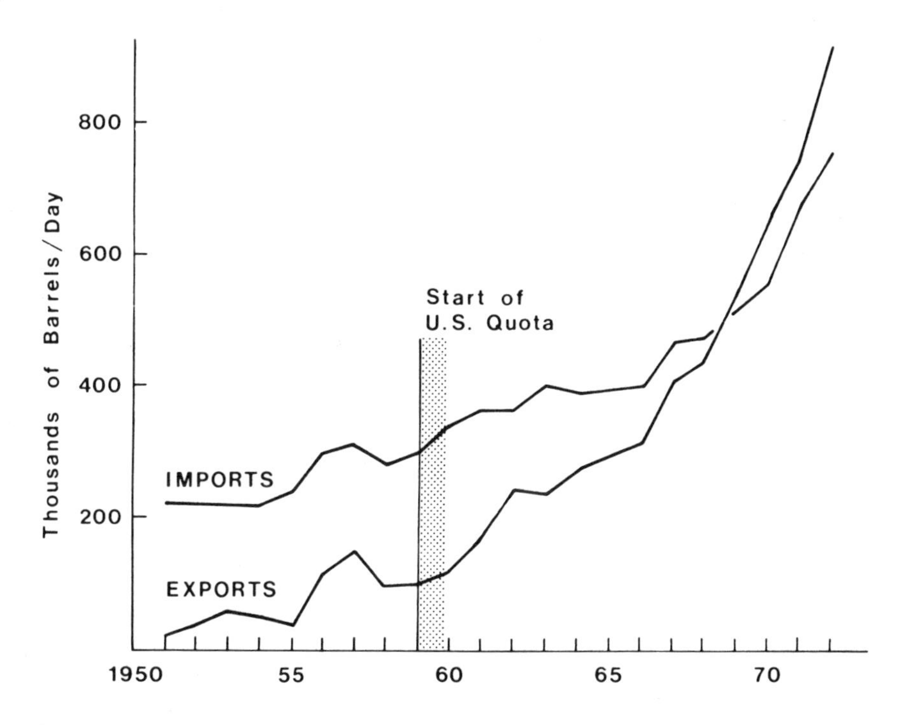

Source: Imports from government of Canada, *Imports by Commodities.* Exports from government of Canada, *Exports by Commodities.*

The integration of the two oil economies, however, contained one major gap: Canada's measures favoring local production were relatively ineffective.[5] Once the overland exemption came into effect, Canadian exports into the United States expanded rapidly, taking advantage of the sheltered American market. Coincidentally, Canadian imports from Latin America and from the

5. Areas west of the Ottawa Valley were reserved for Canadian crude. The enforcement of this provision, however, seems to have been rather lax and its legality questionable.

eastern hemishpere rose sharply.[6] The pattern of these two series is clearly seen in figure 1 (page 11). In 1958, the last year prior to the imposition of the compulsory quota, Canadian imports were some 285,000 b/d and exports (virtually all to the United States) were 87,000 b/d. While exports increased quite steadily to 581,000 b/d in 1969—an increase of 494,000 for the period—imports increased by 244,000 b/d to 529,000 b/d. Prior to 1959, the difference between Canadian imports and exports had been fairly stable (220,000 b/d in 1951 and 198,000 b/d in 1958). So the post-1959 excess of Canadian exports can be attributed largely to the overland exemption. In terms of security, one can say that the 494,000 b/d increase in Canadian exports to the United States yielded a 250,000 b/d net addition to the combined crude-producing capacity of the two countries.[7] In slightly different terminology, the increase of Canadian exports to the United States by one barrel (obtained at the expense of other countries) yielded about 0.54 barrels of capacity to the two countries combined. This number seems sufficiently large to sustain the notion that security was indeed the purpose of the overland exemption.[8]

The increased Canadian dependence on foreign oil first received formal recognition in a 1967 secret agreement between Canada and the United States stipulating that Canadian exports to the United States could increase only at a fixed rate, substantially lower than the trend then current. That was a most curious agreement as, apparently, no enforcement mechanism accompanied the pact and virtually none developed. The Canadians subsequently argued that they had carried out their share of the agreement by imploring exporters to exercise restraint: it was not their fault that they lacked the legal power to control exports.[9]

How were Canadian imports incorporated into the overall quota restriction? First, the overall level of the quota for a given year was set and then the (estimated) Canadian imports were subtracted from it. The

6. We are not referring to shipments of foreign oil to the United States through Canada. The exemption applied only to oil produced in Canada. This was relatively easy to enforce by permitting imports only from Alberta.

7. Two reservations should be made. First, we here equate output with capacity, but in both countries, particularly in the presence of prorationing, the two may diverge. Second, we measure capacity by number of barrels without regard to quality. Since it seems that the quota induced a lower quality crude, the figure of 250,000 is biased upward. We will return, in a more general setting, to the latter question.

8. The years 1970–1972 are omitted in the above comparison because a sharp increase in tanker rates in 1970 substantially reduced the incentive to import Caribbean and eastern hemisphere oil for both Canada and the United States. Also, in 1970, the United States imposed moderately severe quota restrictions on Canadian imports east of the Rockies, which were later relaxed in response to the change in ocean shipping costs. In any event, extending the comparison to 1972 would change the results very little.

9. The agreement, it seems, was not entirely toothless. Seemingly, it slowed down Canadian imports into the Chicago area in exchange for a permit for a Canadian pipeline to go through the United States. See "Clark smokes out secret U.S. Canada oil deal," *Oil and Gas Journal* (March 3, 1969), p. 91; and "U.S. and Canada reveal secret pact details," ibid., March 10, 1969, p. 50.

remainder was allocated to refiners and to importers of record.[10]

Even though Canadian crude was exempted from restriction until 1970, to American refiners, Canadian and domestic crudes were not entirely equivalent.[11] The reason is that refiners earned a share of the quota only according to their input of domestic oil, an extra ten barrels of domestic crude increased the import quota by about one barrel valued at some $1.25. Consequently American crude commanded a premium of some ten to fifteen cents per barrel over similar Canadian crude.[12] The loss of the right to the quota by importers of Canadian crude was not uniform, however, since small refiners earned a proportionately larger share of the quota.[13] Some testable implications derived from the special treatment of Canadian imports are later incorporated in a more general discussion on the allocation of the crude quota.

How were the gains and losses from the Canadian quota distributed? Had Canadian crude production remained constant, the gain to Canadian producers would have been exactly matched by the combined losses to Canadian consumers, who now paid a higher price, and to American importers, who lost some of their share of the quota. As imports from Canada expanded, Canadian producers increased production in response to the demand; some of their gain in revenue realized through higher price was matched by higher real costs, and on that margin the two should have been equal. For that reason, the combined loss to American importers and to Canadian consumers exceeded the gain to Canadian producers. Of course, that extra output reflects a gain in security in terms of increased self-reliance in oil for the United States and Canada combined.

As mentioned earlier, the overland exemption east of the Rockies was replaced in 1970 by an import quota. The new restriction, it seems, came largely in response to the rapid increase in crude imports from Canada and to the associated reduction in the remaining import quota for the refiners of domestic oil.

Unlike the refiners' quota, the new Canadian quota was based almost entirely on historical precedent.[14] The quota for the last nine months of 1971

10. See Appendix A-2.

11. The first major change in import controls undertaken by the Nixon administration, early in 1970, was to impose a strict quota on Canadian exports to the Midwest.

12. An estimate of twenty cents per barrel was made early in 1969. This, however, may contain an upward bias since a damage claim was based on the differential. See *Oil and Gas Journal,* March 3, 1969, p. 91.

13. To further complicate the problem, the "northern tier" refiners had their own special historical program, which is described in some detail by Dam. Trying to convey a sense of the complexity, he writes, "Thus, a decision off-setting an earlier decision was itself partly offset in order to afford protection to particular upper-Midwest producers at the expense of other domestic producers." Dam, "Implementation," p. 33.

14. It is worth noting that following the "secret agreement" American refiners had been under some pressure to voluntarily reduce imports. Those who had complied were now penalized, as their historical base was thereby lowered.

was set as 105 percent of each refiner's imports during the twelve-month period of October 1968 to September 1969, and the percentage was allowed to increase slowly in subsequent periods. An exception to the historical quota was made for refiners in the process of starting operations based on Canadian crude.

Various restrictions, some quite severe, were imposed in special cases such as exchange of the Canadian quota with other oil, eligibility for refiners relying mostly on Canadian oil, refiners who failed to use their entire quota in a given period, and the like. However, toward the middle of 1970, when a sharp increase in tanker shipping rates added charm to Canadian imports, both the Oil Import Administration and the Oil Import Appeals Board responded by relaxing many of the restrictions. This was done on a case-by-case basis, but quite generously, basically allowing Canadian imports to maintain their trend of growth.[15] Indeed, by early 1971 it seems that the main real constraint on Canadian imports was pipeline capacity rather than limited quota.

Mexico. For about a decade, some 30,000 b/d of crude (or unfinished) oil were imported to the United States from Mexico by way of the "Brownsville U-Turn."[16] This devious route involved the shipping of oil by tanker from the Mexican allocation to Brownsville, Texas, where it was unloaded under bond into trucks that then returned south of the border, made their notorious u-turn, and proceeded back to the Brownsville port. The oil which thus qualified under the overland exemption was then shipped to the East Coast. Only in 1971 was this curious constraint removed so that Mexican oil, at it previous volume, could travel directly to its destination.

Although the original overland exemption ostensibly applied equally to Mexico and Canada, the Mexican case differed in that the volume was always firmly restricted. Two points should be noted. First, Mexican oil is more expensive than Venezuelan. Second, all Mexican oil is handled by a single state-run firm—Pemex—that consequently is in a position to extract for itself the value of the quota. The Mexican quota, then, was of little value to American petroleum firms, and we can offer no explanation why the presumably political handout took such an expensive form.

THE SPECIAL TREATMENT OF DISTRICT V

As mentioned earlier, for most quota purposes the United States was divided between one great bloc known as Districts I–IV and a second area,

15. One cannot fail to be impressed by the flexibility of the defense requirements.

16. The period followed an agreement signed in 1961 between U.S. authorities and Pemex, the Mexican state-owned oil monopoly. See Edward H. Shaffer, *The Oil Import Program of the United States.*

District V, which comprised the Pacific states, Arizona, and Nevada. Because these two major regions constitute largely separate markets, the setting up of differing controls seemed logical. But, given that national defense was the main purpose of the quota, one would have expected the two to be treated consistently in terms of self sufficiency, health of the industry, and so on.

Actually, their treatment was widely divergent. In the beginning, the voluntary quota did not apply at all to District V; when it was initiated there, early in 1958,[17] the level was set most generously at 220,000 b/d whereas actual imports that month were only 151,600 b/d.[18] With the inception of the mandatory program, enforcement continued to be less rigid in District V: imports provided 20 percent of the total crude used in that area as compared with about 10 percent in the eastern district.[19] It appears also that the value of import tickets was lower in District V, where occasional references place it at eighty cents per barrel as against the commonly asserted value of $1.25 per barrel in the rest of the country. Finally, when in the early seventies Canadian crude was subjected to a quota, this applied to Districts I–IV, but not to V.

Intuition suggests that the difference in treatment resulted from the relative political muscle of crude producers in the two areas. In the three major crude-producing states in Districts I–IV—Texas, Louisiana, and Oklahoma—as well as in many others there, crude output is prorationed. Such is not the case in California. In addition, unlike these states, California is not a member of the Compact, a congressionally blessed organization of states designed to coordinate their oil-oriented activities. Whether this hunch is correct must remain for the time being an unanswered question. By the time the hypothesis was formed, no independent data remained (to our knowledge) by which it could be tested.

---

17. See Dam, "Implementation," fn. 86.
18. *Oil and Gas Journal,* May 5, 1958, p. 80
19. A similar disparity, however, already existed during the voluntary quota period.

2 • A Voluntary (?) Quota

The relationship between economic activity and its surrounding legal framework has recently received increased attention in the professions of law and economics. This new, or rather revived, interest promises significant advances in the development of both empirical and theoretical economics. Understanding of the legal aspects of a problem provides the investigator with a set of constraints to be used in generating predictions, while changes in the economic environment lead to changes in the legal framework.

Although the stability of our legal institutions is felt by some to rest on a pinnacle of ethics, the argument can also be advanced that laws are constant because it is economically efficient to have them so. If the legal system is unduly vulnerable to change, then resources that might better be employed for production will be spent on inducing and adjusting to such alterations. The United States is said to provide a set of stable legal constraints—a system of rule rather than authority—under which neatly specified rules of the game preclude such wasteful expenditure.

It may appear that the oil import quota offers a pertinent case for the study of this interaction of law and economics. However, once one starts to probe into observed behavior under the quota, the role of the legal system very quickly recedes to provide, at best, a background.

The starting date of the quota presented its first legal ambiguity. Although the mandatory quota began only in March 1959 by presidential proclamation, it had been preceded by spotty compliance with a series of voluntary quota systems dating back almost four years. In brief, as early as July of 1954, President Eisenhower established an Advisory Committee on Energy Supplies and Policy Resources, and by the following February, the committee had recommended a voluntary reduction in oil imports back to this 1954 level. Subsequently, the director of the Office of Defense Mobilization was assigned to oversee the industry's compliance.

Following the failure of this program to hold imports to the earlier level, a Special Committee to Investigate Crude Oil Imports was established in April of 1957. By July, it recommended that imports east of the Rockies be based on domestic production, setting a 12 percent ratio of imports to domestic production as the desirable relationship. Strangely enough this 12 percent relationship was later to become fixed by the mandatory program, but only after a lower percentage was found to be consistently overrun. The Department of Interior was assigned to oversee the newly revised voluntary system.[1]

Not everyone was satisfied with the voluntary arrangement, and opponents regularly and loudly proclaimed their objections. Oil interests were not the sole vocalists: foreign governments—particularly Venezuela and Canada—along with coal interests and large oil consumers, also contributed. Their arguments for and against the voluntary quota (particularly the latter) were voluminous, as were their views regarding the effectiveness of the system.

Some of the debaters took direct action in expressing their concerns. In 1958, for instance, Delta Refining Company and Superior Oil, apparently deciding that they had volunteered enough, began importing without first obtaining the blessing of a required voluntary allocation. Delta's and Superior's aberrant behavior aroused criticism, as evidenced in a statement by the voluntary quota administrator, Captain Carson:

> If Delta Refining Co. and Superior Oil continue with such an agreement, and imported crude oil is being landed in New Orleans without a government-approved import allocation, these two companies will have to share the responsibility for an action which may tend to impair the adequacy of a program designed to enable our domestic oil industry to meet the requirement placed upon it by consideration of national security.

The administrator later pointed out, however,

> . . . that in the event that Delta continues in their pursuit, only the weight of public opinion could be brought to bear against them.[2]

This was only one of many instances where the voluntary arrangement failed to accomplish its nominal objectives. Such flaws were seized upon by groups intent upon attacking either the entire program or at least the enforcement of the existing quota. If this voluntary system is judged only by a comparison of its stated objectives with the official record of imports, its critics are certainly correct in saying that the program failed to meet its objectives totally. However, if performance is measured in other dimensions, the record is more favorable.

The proponents of increased control—i.e., particularly domestic producers and coal interests—argued for, at the very least, a more comprehensive quota to encompass refined oil products as well as crude oil. During the 1957-1959

---

1. For additional information, see Aubrey M. Kirby, Jr., "The Chemical Industry and the Oil Import Program."
2. *Oil and Gas Journal,* January 22, 1958, p. 124.

period, imports of refined products increased substantially, both in absolute and relative importance. This would not have been surprising under an effective crude mandatory quota making the import of close substitutes worthwhile. The fact that it is observed under voluntary conditions, however, speaks a reasonable obedience to the rules. Additionally, 1954 was again proposed as the model period in the sense that the loudest proquota groups also sought a reduction, on a mandatory basis, to levels of crude oil imports below the prevailing voluntary levels, back to the 1954 ratio.

In the same sense that the term *oil* or even *crude oil* may be misleading by its apparent simplicity, so too is *the quota.* Both oil and quota definitions are in fact so infinitely variable that to conceive of these terms as describing a unique state is extremely naive. The inherent complexity of the actual quota, be it designated as mandatory or voluntary, leads to an immediate rejection of the standard picture, in which monolithic oil interests maintain impenetrable solidarity. Comrades on a particular quota provision were opponents on others, and this schizophrenia prevailed in many cases within individual companies whose interests encompassed several affected areas. Each of the many modifications of the quotas redistributed gains and losses, and because of this variety of effects, it simply is nonobvious what distinction is involved in the voluntary-mandatory dichotomy.

Any particular specification of "the quota" (i.e., the quota as it is actually enforced) implies a wide spectrum of gains and costs: no two individuals are in identical positions simply because of the tremendously varied components of what is loosely called the oil industry. To mention a few dimensions by which firms differ: the fraction of total crude output from abroad; the ease of importing Canadian oil; the share of the heating oils market in the Northeast; the extent of acceptable pollution-producing crudes. Any quota system necessarily affects participants differentially along any of these lines as well as along many others. The variants obviously expand, once other interest groups such as coal, shipping, and chemicals are added.

This complexity creates problems of enforcement. It also, however, allows owners of the quota to adjust its provisions to screen out would-be free riders. In the theoretical presentation, we attempt to show that in this situation the rising supply is of *rights* to the quota; therefore, in the general case there will always be intramarginal gainers as well as the marginal man. Each respecification of the quota rules implies a different ordering of the individuals competing for quota income and thus, in a sense, provides a mechanism for automatically precluding participants from receiving gains without paying correspondingly. This would be the case if the costs of adjusting the regulations, and of accurately forecasting the effects of each adjustment, were zero. Such costs complicate the problem of taxing all owners for their share of the quota, but this is a general problem affecting any good, since it is never possible that every marginal condition is satisfied.

Not surprisingly, then, mandatory quota fans were hardly unanimous or unwavering in their stances. The Independent Producers Association of America (IPAA), for example, first publicly supported the return to 1954 proposal but later endorsed the Ikard bill which would have auctioned off import rights rather than assigning them.[3] Of course, these groups may have been motivated by altruistic national defense interests and thus constituted a voluntary army of sorts. In any event, as the representative body for domestic producers, the IPAA was little perturbed by the auction aspects of the Ikard proposal, since the main alternatives included allocation of import rights to refiners and to established importers. Many such proposals were advanced, each claiming superiority in serving the national interests.

Captain Carson was repeatedly drawn into public defense of the effectiveness of the voluntary program. On one such occasion he assessed its results up to May of 1958:

> Carson said the program has achieved 95% compliance because now only 2 of 58 importers have failed to signify compliance, and one of the remainder may comply soon.
>
> Actual crude import into Districts 1 through 4 from July through March, Carson said, have been 17.4% below what the importers had originally scheduled to bring in, and have been only 2.3% above goal. On this basis the program has been 97.7% effective.
>
> As originally scheduled, these imports would have been 16% of domestic production, but from July through March they have been 13% and September through March they have been 12% or 0.5% above the goal.[4]

For whatever reason, whether or not it was a doubt of Carson's ability to discern the intentions of oil importers and subsequently to influence their behavior, his plea to maintain the voluntary quota was greeted in an IPAA meeting by a unanimous vote to endure a mandatory system. Apparently this reaction to Carson's assessment was shared by the Texas Independent Producers and Royalty Owners. Following a similar presentation, they too unanimously voted to support the Ikard proposal, though they courteously added an acknowledgment of Captain Carson's "sincere and outstanding efforts."

In some respects, both sides of the mandatory-voluntary struggle were correct in their appraisals of the current system's effectiveness. By their own admission, importers were not strictly adhering to their quotas, and in some cases, these indiscretions were sizable. Since available penalties were unlikely to apply uniformly and since unsanctioned increases of imports might earn

3. Ibid., April 7, 1958, p. 101.
4. Ibid., May 19, 1958, p. 99.

future higher quotas if the historical basis were subsequently to be chosen (which eventually proved correct), the irregularities by some companies should not be surprising.

Two pieces of evidence that support the claim of effectiveness of the voluntary system went understandably unmentioned during this debate. Later, we argue that an effective quota will induce a shift in import composition toward higher quality oils; when these are exempt from the quota, as they were under the voluntary system, their share of total "oil" imports will increase even more. Actually, both the relative share and the absolute size of oil-product imports increased abruptly with the quota's inception, as subsequent attempts to include them in the restrictions witness.

A more curious bit of evidence supporting Carson's contention concerns the exchange of import rights. Suppose no resources of any kind had been available for enforcing the quota. In such a case, violations would have abounded. That being the case, why would any holder of import rights have chosen to take part in this trading activity? Reports of the value of these rights are sketchy and must be considered as rough estimates of their actual magnitude. Taken at face value, they indicate a price of approximately eighty cents per barrel on the East Coast, or some two-thirds of the corresponding price under the mandatory program (which included oil products). Apparently some participants were acting with noticeable self-restraint.

A profit-maximizing firm would not have been inclined to restrict its operations simply at the request of a government official: only the threat of some penalty would have impelled reluctant volunteers to accede. Consequently, the existence of penalties must be inferred from observed changes in behavior such as the trading of rights. The very absence of evidence on the form of such penalties strongly indicates a case of authority rather than of rules, the unspecified deterrent apparently resting in administrative hands.[5] It is not suprising, therefore, that Captain Carson did not use this evidence in support of his contention.

Even when the quota's voluntary status gave way to a mandatory program, the governmental enforcement mechanism was never one of its prominent features. Nevertheless, rights to quota income were established and so too, it seems, were rights over the voluntary quota. In the policing section we argue that even the mandatory program was largely self-enforced, and our observations on the voluntary programs, though sketchy, are not inconsistent with this hypothesis.

The designation of voluntary as opposed to mandatory to some degree obscures the nature of the constraints involved. Both quotas were voluntary in

5. One form of persuasion is the withholding of general government business from noncompliers. "Bidders for new contracts must agree to stay within quota. . . . The Military Petroleum Supply Agency has taken steps to *enforce* President Eisenhower's *voluntary* import program." (Italics added.) Ibid., April 21, 1958, p. 105.

the strict sense, since anyone could become a legal importer of oil by acquiring rights by methods ranging from the outright purchase of qualifying companies to the exploitation of the inevitable loopholes in each program. The voluntary-mandatory distinction more aptly applies to the method and degree of enforcement of the programs, but even here, substantial evidence indicates that the change to mandatory status did not alter the essential aspects of the enforcement mechanism. The program remained a child of authority, and it is the actions of this authority that we attempt to describe and analyze in the following sections.

Any thorough analysis of the constraints facing maximizing quota owners and those excluded by the quota is complicated by the difficulty of discovering a set of regulations that fully defines property rights to the quota. The enigma poses problems both for individuals affected by the quota and for those interested in explaining their behavior. For the players, the complexity suggests that the quota's enforcement was entrusted to administrative hands to counter the appearance of unanticipated loopholes and other problems. Such intrinsic ambiguity presages both the dissipation of resources (as individuals seek to capture portions of the nonexclusive income) and modifications of the rules of the quota itself in response.

The switch to mandatory quotas clearly did not bring the first specification of rights, but it did serve to strengthen them, i.e., to increase their value. One intent of the following sections, particularly that dealing with petrochemicals, is to point out thc difficulty of clearly specifying rights, even when there is apparent explicit legal approval. In this case, resources were consumed in preliminary attempts to obtain favorable status and, subsequently, to exploit loopholes in the regulations.

The concluding petrochemicals sections on Puerto Rico and the Virgin Islands pick out two examples of discretionary aspects of the quota. These experiences were selected not because they were unique in concept but, rather, because of the size of the wealth transfers involved and the profusion of arguments they generated, and because of the clarity with which they illustrate the authority-versus-rule argument.

3 • Theoretical Framework

The simplest model of an import quota is portrayed in figure 2 (page 23). D and S_D are the respective domestic demand and supply of a commodity. S_F is the (perfectly elastic) foreign supply, and foreign price is P_F. A quota Q_F is granted so that $S_D + Q_F$ is, for prices above P_F, the combined supply to the domestic market. Q_E and P_E are the post-quota equilibrium quantity and price. Of the final quantity, $Q_E - Q_F$ is provided by domestic suppliers and Q_F is obtained from foreign sources.

The quota's national-defense objective of increased self-sufficiency could be attained alternatively via a comparable inducement to local supply coupled with a disincentive to importation. One such possibility would be a tariff that would also result in price P_E. To domestic producers the two would generate the same benefits, which is the area *(A)* between P_E and P_F left of S_D. In both cases the trade restriction implicitly, but automatically, grants domestic producers property rights over that gain. In this respect, also, the two are equivalent.

So far, our stick-person view of quotas has achieved a certain desired, though costly, simplicity. The price of out innocence is that we must completely exclude all effects of potential rents on resource allocation. For example, logical consistency forces a stringent and improbable interpretation of *quota* in this framework, since all individuals are assumed to obtain costless information as to how the quota is defined and enforced. In addition, the quota must somehow appear without forewarning and operate without the consumption of resources.[1] Once in effect, the quota itself can have no effects on the cost functions of its participants.

Looking again at our diagram, Area A is the increase in domestic producers' income attributable to the costless enforcement of quota Q_F if their production costs are unchanged by the quota. Since this group has the right and the capability to capture this rent without cost, maximization assures us that they will exploit the opportunity to the fullest extent. Area B is the remaining rent said to exist, but there is no explanation of how it drops

1. If this were in fact the case, the change in present values of quota recipients would have caused a sufficient jump in their stock prices at its inception. No such phenomenon was observed.

into the pockets of our quota participants. Unlike the gain from additional domestic oil sales, the rules by which *B* can be captured are not spelled out. At least, they were not included with the instructions for assembling this diagram.

FIGURE 2

Simple Model of an Import Quota

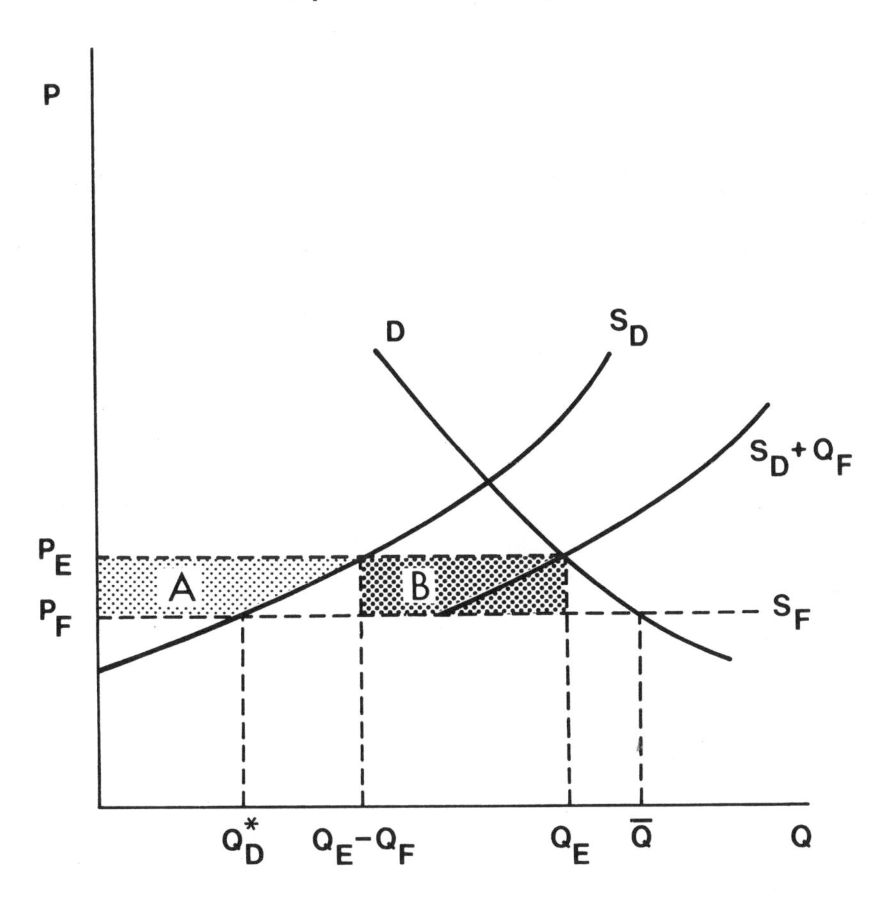

Imagine, for instance, that an advertisement were to appear offering rights to import oil at a zero price, while they last, at your neighborhood Oil Import Administration office. With one of these rights and two dollars you could have delivered to your home or refinery one barrel of first class Kuwait crude. Once delivered, this barrel of oil would sell for $1.25 more than you paid for it including the transportation charges from the Mideast. Due to limited supply, the ad goes on to say, the number of tickets will be allocated in books of ten tickets, via the first-come-stand-orderly-in-line method. Predictably, hopeful candidates will appear prior to the opening of the window for their share of *B*. Since a book of ten rights is worth \$12.50, each one who waits in line will spend up to \$12.50 worth of time. If the hourly wage of our identical would-be importers is \$6.25 per hour, the line will form two hours prior to opening, and if Q_F (the total allocation of rights) is 100 barrels, the entire area *B* (i.e., \$125) will be obtained by ten people who spend a total of twenty hours waiting for tickets.

With the quota income rights and waiting cost so defined, the net value of the quota reduced from Area *A* + *B* to Area *A*; but why are the recipients of *A* so lucky? If there are competitors for *A*, as there were for rights to *B*, must not the right to collect *A* similarly be endangered by competition? The potential recipients of *A* would also be willing to spend resources—this time up to an amount equal in value to *A*—in order to obtain the quota, Q_F. Let us suppose a notice similar to the import-rights advertisement is posted in officialdom (with the identical waiting rules), but the offering this time is the right to establish a quota of Q_F. In other words, the right entitles one fortunate oil lobbyist to have enacted a quota potentially generating Area *A* to domestic producers.[2] The recipient must, however, be the first one to the window since only one such right is being offered. Following our earlier story, *A* ends up being transferred to the first individual, who in turn spends it from his spot behind the import-quota-rights window.

Our selection of the waiting method for allocation of quota rights is purely expository. Many alternative arrangements could be used to assign the rights to these quota rents, but regardless of the actual system used, the behavior of competitors will always serve to reduce rents. We doubt the method of dissipation would, in fact, be similar to our waiting example. Greater imagination might have individuals arm wrestling for the rights.

The waiting people in our example had offered to them a clear and restricted choice that simplified the storytelling. This choice can be viewed as fully parallel to any other production decision, in the sense that the extra hours spent standing at the ticket window necessarily excluded an opportunity to work at other activities that would reward them with \$6.25 for each hour. Consequently, each waiting person will apply resources to obtain quota income until the value per hour to him just equals in value an hour spent, for instance, producing domestic oil.

So far we have dealt only with two polar cases: (1) the textbook example, where both areas, *A* and *B*, are captured; and (2) the case where the entire gain

2. Here we ignore the B rents, since if the rights were structured as proposed, both of these areas would be up for grabs. In addition, in order to obtain this equilibrium solution, the winning individual must be able to costlessly collect from the domestic producers the entire value of A. Again, this is not a derived result inasmuch as we asserted the allocation criteria and pricing constraints.

is dissipated. The first case cannot be of any significant interest empirically, since only an unnoticed and immovable deity could establish and operate such a quota. Without the help of this cooperative, though deaf, spirit, we would once again be pursued by the devils of inconsistency. How could it be, for instance, that individuals affected by the quota could costlessly obtain information about these complex regulations, establish (again without incurring expense) an absolutely effective enforcement mechanism, and yet still find themselves unable to exploit the potential gains from exchange found in the so-called welfare triangle? This result cannot be derived from known economic principles because of the seeming impossibility of specifying transaction-cost constraints capable of implying this, and only this, behavior. In effect, this outcome simply must be asserted.

The second case leads us to examine other features of the quota, though it is perhaps becoming obvious that *the* quota is something other than a simple limiting of imports. For a wealth-maximizing individual, any restriction placed on his right to import is, in and of itself, irrelevant. His real concern is the wealth change induced by the quota, and to discover how this will occur demands a complete specification of the de facto rights to rents from the quota.

This complication brings us back to an earlier restriction imposed by figure 2. The actual quota assigned the rights to income in a manner that was hardly neutral with respect to the cost functions of our domestic producers. Later, in more detail, we argue that the quota would be expected to lead to a substitution in the types of oils produced both domestically and abroad, causing the cost functions to shift upward. This adjustment can be likened to our earlier discussion of the assignment of rights, since it results from the producers' right to adjust, within limits, the composition of their output. Additionally, once we allow for the costs of enforcing the quota, this result is automatically achieved, since positive enforcement costs assure that it will never pay to completely prohibit substitution of oils.

The polarity of the second case is equally unlikely to be the general result, simply because not all individuals are identical with respect to their costs of standing in line or, more generally, their comparative advantage in obtaining rights. When some individuals have lower costs than other (i.e., when the supply of waiting persons is upward sloping), the entire quota rent will not be spent waiting, though the marginal condition necessarily holds. For instance, though one lower-cost waiting individual would be willing to arrive ten hours prior to the opening of the window, he will still be required to stand for only two hours.[3] This individual will capture some of the quota rents (eight hours' worth), and with this situation of rising supply, all but the marginal waiting person will be able to pocket some of the quota rent.

3. For simplicity the marginal waiter's time is assumed to be the same in both cases.

We have tried to show that the gains from a trade restriction will fall in quite different directions depending upon the particular structure of the quota or tariff, even in our extremely simple case with very limited adjustments possible. Because of these variations, potential gainers and losers will behave differently in expending resources to maneuver toward one or the other system prior to its establishment.

The resource cost of obtaining a tariff, however, is not necessarily the same as that for obtaining a quota. Nor is the resource cost of policing necessarily the same in the two cases. The gain to domestic producers is obtained, then, at some cost, the magnitude of which will vary with the form of the restriction. The net gain therefore, is presumably not the same for a tariff as for a quota.

Perhaps the widest divergence between the two systems relates to the (dis)incentives offered to importers. The tariff imposes a clear tax on them, whereas a quota will always provide some intramarginal gains (the rising supply situation). Support from these hopeful beneficiaries may explain why a quota was actually selected instead of a tariff.

In the tariff case, the amount paid is a cost to market participants, as is the so-called welfare triangle. The hypothesis of maximization implies that both of these costs, as viewed by the market, will be minimized by importers who will exploit other margins in addition to quantity, since the costs of enforcing either the quota or the tariff are positive.[4] Adjustments at these margins will create new inefficiencies, but these are accepted in return for the reduced total cost of tariff plus welfare triangle.[5]

In the case of a quota, the situation is more complex. To begin with, the mere knowledge that a quota is imposed does not indicate who will receive the right to the transferred wealth.[6] In general, as with a tariff, we would expect a minimization of inefficiencies subject to the constraints actually imposed. To the extent that property rights to gains from the quota are perfectly defined, the minimization process applies only to welfare costs. On the other hand, the quota may be granted to nobody in particular, in which case resources will be spent in acquiring rights to it. The quantity to be minimized is then the amount of these resources plus the welfare triangle. It seems clear that whichever way property rights to the quota are defined, adjustments will differ from those made in the tariff case.

One crucial difference between a tariff and a quota concerns the allocation criterion itself. In the case of a tariff, one has to pay at a point and time designated by custom authorities, and there is some (perhaps very small) cost

4. The tax, if costlessly administered, is not strictly a cost to society, though the donors might disagree.

5. See Barzel, "An Alternative Approach to the Analysis of Taxation."

6. An implicit, but rather important, additional assumption in measuring the transfer is that it is too costly for the quota owners to price-discriminate. Had this not been the case, Q, the prequota equilibrium quantity, would also be the postquota quantity, since it provides the largest potential gain. When transaction costs are zero, a potential gain will be converted into an actual gain.

associated with the act of paying. For a quota, the exact criterion for allocation will differ from one instance to another.[7] In the case of crude oil allocation, each firm was given the option of adopting either its historical importing base or a domestic refining base. Property rights to the quota were not thereby perfectly defined. A firm had to spend resources to demonstrate its eligibility; more importantly, its expenditure of resources could modify the amount earned, particularly under the refining criterion. On that basis, the marginal condition for maximizing wealth obviously must take into account the opportunity to earn import rights by increasing the stated quantity of domestic oil processed. Thus, the new condition will be to process oil to the point where the marginal value of one barrel equals its marginal cost minus the value of the extra import right obtained.

The actual allocation, mostly to refiners and historical importers, was based on a complicated formula. How were these specific criteria reached? Surely potential claimants to the quota spent resources to divert the quota in their direction, and in each case, the extent of the expenditures would have been equated with expected gain on the margin.[8] Available data seem insufficient to test hypotheses regarding either the total resources used for such purposes or this assumption of marginal equality. We will, however, later bring to bear qualitative data indicating that resources were indeed used for such purposes. The above analysis, brief as it is, evinces that an important effect on resource allocation is missed in the conventional analysis of welfare loss due to the quota[9] as represented solely by the welfare triangle.[10] Namely: how large is this total resource use?

From traditional analysis it might appear that the only bone of contention is who will (costlessly) get the quantity Q_F and thereby gain P_E - P_F per barrel. In figure 2 it was implicitly assumed that we know precisely what constitutes domestic and foreign oil, that oil is a homogeneous commodity, that the regulations are fully and costlessly enforced, and so forth.

Consider now figure 3 (page 28), where alternative quota allocations (in barrels per period), *T,* are measured along the horizontal axis and the value per barrel, *V,* is measured along the vertical axis. Given a quota Q_F, the value of a ticket—i.e., the right to an import quota of one barrel of foreign oil—is V_0. Its magnitude is essentially determined from the total allocation Q_F, and is the difference between the price of domestic oil, P_E, and that of foreign oil, P_F

7. To some degree, a similar consideration may apply also to a tariff.

8. In addition, before the quota became fixed, eventual losers had an incentive to spend resources to try to avert the quota or at least to reduce its severity.

9. Recently the problem has been recognized variously by Tullock ("The Welfare Costs of Tariffs, Monopolies and Theft"), by Yoram Barzel ("A Theory of Rationing by Waiting"), by Steven N. S. Cheung ("A Theory of Price Control"), and by A. O. Kreuger ("The Political Economy of the Rent Seeking Society"). Levis Kochin argues convincingly the Adam Smith was well aware of the problem but that it ceased to be recognized around the time of Marshall.

10. In Figure 2 (page 23), the triangle is contained from below by S_F and from above by D and by $S_D + Q_F$.

FIGURE 3

Value of a Ticket to Import Oil under the Quota

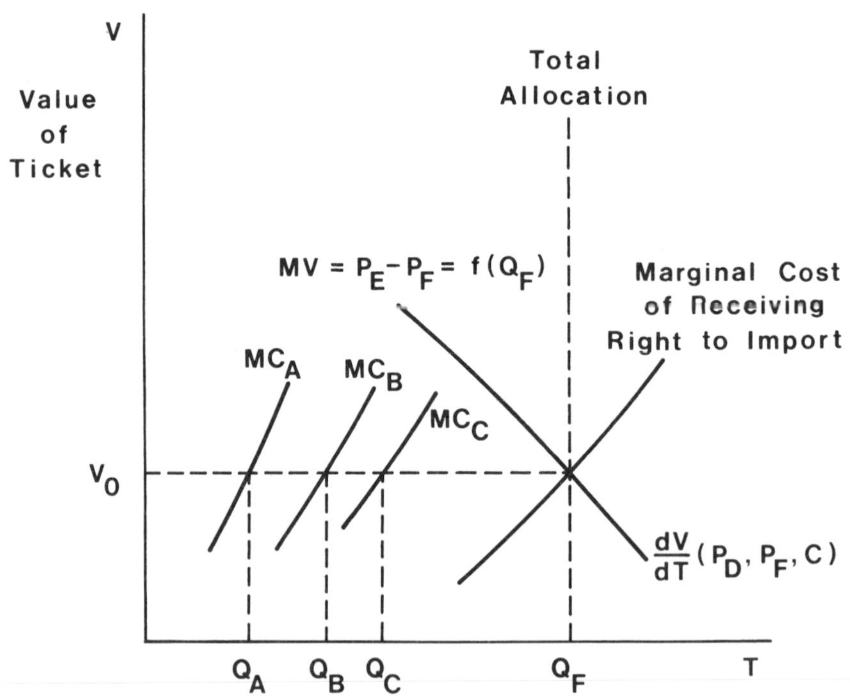

(see figure 2). Q_A, Q_B, and Q_C are the ticket allocations to individual firms such that $\Sigma Q_i = Q_F$. Q_A is the quota level chosen by Firm *A*. By changing the amount of domestic oil it refines, or by taking some comparable action, Firm *A* can affect the amount allocated. MC_A is the marginal cost curve for tickets to Firm *A*. Q_A is the quantity it desires when their value is V_O. For the ticket market to reach equilibrium, the value of a ticket must be such that the available quota is also the sum of individual quotas desired by firms in the industry.

The marginal cost function is really composed of many cost functions reflecting the various costs of adjustment margins, where each point is the minimum cost of obtaining the right to an extra barrel. Also taken into account are the costs imposed by other firms adjusting to the same forces (i.e., this is a Cournot-type solution).

Earlier, we restricted costs of obtaining quota rights to those incurred by standing in line at a fictitious ticket window, and the marginal cost functions presented here can be likened to our earlier situation of rising supply. In this case, however, many avenues may be open for obtaining quota rights.

The preceding has treated crude oil as a homogeneous commodity. In reality, at any point in time and essentially within the same market, crude oil varies greatly in price. Lighter crudes normally fetch higher prices than heavier, sometimes twice as high. While it is difficult to incorporate such quality differences in the above diagrams, the quota will obviously induce a switch in the composition of imports toward the higher price crudes, thus changing the ticket values. This hypothesis, together with its corollary that domestic production will shift toward heavier crudes, will shortly be presented in greater detail and then tested.

In the following pages, we attempt to describe this adjustment process and to delineate, when possible, the nature and magnitude of the associated costs. We argue, however, that because quota regulation involves much more than mere quantities, resources will be spent in advance of the quota to influence this multitude of margins. At best, not all ambiguities can possibly be eliminated from the quota specifications, and market participants will adjust their resource use to take advantage of remaining loopholes.

This acknowledged proliferation of effects on resource allocation does not, we argue, mean a *larger* total effect than if quantity were the only subject of competition among market participants. On the contrary, the sum total of these effects constitutes the minimization of the loss. Only under the most adverse conditions will the entire area of quantity Q between P_E and P_F be dissipated. Various adjustments along all the available margins will result in the minimization of that dissipation.[11]

That is, for any given level of contractual cost, participants will choose the arrangement that entails the smallest intramarginal loss. Suppose, for instance, that two commodities are being considered for restriction, and that only one will be actually controlled. For further simplification, let both commodities be produced under identical and perfectly elastic cost conditions, and let the first commodity's demand elasticity approach infinity while the second's approaches perfect inelasticity. The application of a quota that succeeds in producing identical price increases will produce the smaller loss in the case of less elastic demand.

In a manner similar to the tariff example, market participants will act to minimize the damage done by the quota. However, in this case the sum minimized will extend conventional welfare loss to include the resource cost of obtaining and enforcing the quota. With respect to these latter costs, the same elasticity consideration holds.

To conclude this section, a word on the transaction cost constraint may be in order. Maximizing individuals would completely avoid any losses imposed by the quota if exchanges between affected individuals could be arranged

11. See Cheung, "Price Control," and Barzel, "An Alternative Approach to the Analysis of Taxation."

without cost.[12] Once the various rights to action were delineated, the intramarginal gains would then be redistributed but the marginal condition existing prior to the quota's inception would be retained, thus creating the largest possible pie to be sliced. In other words, no actual limiting of imports would be observed. However, once we acknowledge that these devices to do away with dissipation do themselves involve real costs, we can understand why the concerned individuals were willing to tolerate some inefficiencies. Throughout the remainder of this book we attempt to show how some of these adjustments to the quota actually took place within the oil industry.

12. This is simply another statement of the Coase theorum which states that in the absense of transaction costs, all potential gains from exchange will by realized. See Ronald H. Coase ("The Problem of Social Cost").

4 • The Policing of the Quota

The official monitoring agencies of the mandatory program were the Oil Import Administration (OIA) and its adjunct, the Oil Import Appeal Board (OIAB). The latter implemented the announced policies of the OIA and higher agencies by reponding to petitions from aspiring importers. Like any umpire, the OIAB continually aroused debate, and industry publications heatedly chronicled its behavior.

> The Oil Import Appeal Board not long ago had the well-deserved reputation as an agency with a heart of stone.
>
> It not only had precious little oil to give away in "exceptional hardship" cases arising under mandatory oil-import controls. Members were rarely moved by even the saddest of stories which a hardened reporter could scarcely read without weeping.
>
> But no more. Since last year the atmosphere has changed completely. OIAB not only has more oil to give away. It also has a much lower threshold of sympathy.[1]

Before this awakening of humanitarian concern, the Appeal Board occupied itself with the attempts of individuals to expand their share of the quota or, rather, its concern was to hold any such alterations to a minimum. The extremely limited number of tickets issued to the board by the OIA made such control easy early in the game. At the outset of the mandatory program, several minor quotas were issued to companies whose clear right to such an allocation had simply been overlooked while the paper was being shuffled. A thornier problem arose where companies appealed for expanded importing rights on the grounds of undue hardship. Individuals "held captive" by a market, who felt they were being held over a barrel, frozen, punished, or generally abused, were allowed to petition for a larger piece of the quota. However, the board was not empowered to grant new allocations that were fundamentally at odds with the guidelines for the program itself.[2]

---

1. "Watching Washington," *Oil and Gas Journal,* October 7, 1968, p. 85.

2. In Industrial Solvents Corporation's appeal, the board said that it would only modify allocations made by the Administration—i.e., it had no authority to grant allocations on the basis of eligibility (as defined by the OIA). *Oil Import Digest,* Case No. T-5, p. B-51.

A number of suits challenging the legality of the board and of the entire program were filed (or threatened), but incredibly none ever found its way to a legal decision. Some complaints were withdrawn after the board's reexamination led to more favorable action; others seemed simply to fade away.

The first full-scale attack on the legality of the import program was presented in 1959 by Texas-American Asphalt Co. (T-A). After obtaining a temporary injunction from a district court to restrain U.S. Custom officials from stopping a shipment of crude oil imported from Venezuela, T-A went to court seeking to bar the federal government permanently from stopping their crude imports.

Rallying to the aid of the government as friends of the court were the Independent Producers Association of America and the Texas Independent Producers and Royalty Owners Association. These groups defended the stability and health of the independent domestic producers of crude oil, particularly those in Texas. They professed a total lack of understanding of T-A's selfish attitude or of any reason for constructing new plants designed to process foreign crude. T-A's actions, they said, were totally unwarranted in light of the import control program and the "widely-publicized mandate from Congress" to strengthen the entire import control mechanism.[3]

If Congress had issued such a mandate it never appeared in writing: neither the board nor any part of the entire mandatory program was the creature of specific legislation. Nevertheless, T-A agreed to seek relief through the Appeals Board, and this proved to be a wise decision when an allocation was subsequently granted despite the company's earlier failure to meet more obvious eligibility requirements.

We are unable to explain why the T-A example was not followed by a host of others who could recognize a similar inducement to withdraw complaints, except perhaps that going to the courts is a costly activity. Further, serious legal opposition to the import program may have been tempered by a steady stream of innovative suggestions offered by participants in the program. Whenever some aspect of the allocation criteria proved to be ambiguous or particularly unpopular, this creativity emerged.

One of the first such flashes of inspiration appeared after the presidential proclamation in December 1959, allowing the board to grant new quotas for oil products. Hardship was again a deciding factor in these allocations, as the board was empowered—within its small capability—to grant allocations to anyone who could show deprivation due to a lack of imports recorded during 1957 (the base year).

This early portion of the board's life had been a period of relative tranquility, but as time passed, its powers of allocation increased. In addition

3. See *Oil and Gas Journal*, June 29, 1959, p. 60; and *Oil Import Digest*, Case No. 59, 1959, B-12.

to its kitty of tickets to be awarded on a discretionary basis, the board acquired the ability to ease, or to aggravate, such problems of importers as late arrivals of tankers, confusion over ticket eligibility, or many of the other complexities in day-to-day operation under the quota. These arm-twisting capabilities of the board altered the costs faced by participating companies, and we hypothesize that this power was used to help enforce members' rights to quota income.

Although the vast majority of appeals to the board were based upon hardship, nowhere did we discover a delineation of this criterion. Judging from the low success rate of applicants, it appears that they, too, failed to grasp the full meaning of hardship. This ambiguity certainly added to the discretionary power of the board which, we argue, is consistent with the hypothesis that the agency played more the role of policeman than of nursemaid.

The potential gains that we have postulated to result from the adoption of a quota accrue to domestic producers and to anyone who ends up with the right to import oil.[4] Because this gain was created by a reduction in the market quantity (discriminatory pricing thus being expressly excluded), any individual in the cartel has an incentive to expand his right to import. Unfortunately for the remaining members, this gain can be realized only at the expense of a reduced value for their part of the quota.

If the OIA and OIAB are thought of as part of a mechanism for enforcing the property rights of domestic producers and of importers to the quota income, they would be expected to act to maximize the value of the entire quota, subject to the constraints that implied its initial adoption. The mass granting of additional or expanded allocations would necessarily refute this hypothesis, and this was not observed. The right to import crude oil was not significantly altered during the early life of the program, but major changes were made almost immediately in the allocation of rights to import oil products.

The actual enforcement of the oil import program cannot be fully described as the actions of responsible agencies carrying out the letter of the quota regulations, since that letter was sadly ambiguous. The goal of a maximizing owner of the quota, on the other hand, would be the enforcement of only such regulations as would establish, de facto, his property rights in it.

An importer forbidden to use any form of price discrimination could attain a positive ticket price only by inducing a reduction in the aggregate quantity of imported oils. The obvious role of the OIA and OIAB in this context was to effect this result, but the method for achieving it was less clear. No set of regulations can entirely eliminate rent-reducing behavior. This is particularly

---

4. That is, so long as the cost of exercising this right is not prohibitive.

true if no exogenuous resources are available to pay for enforcement at the time the regulations are set up.

Actions that reduce the rent to quota owners are of two general forms: those that reduce both the quota's value and its welfare loss by increasing the total quantity imported; and those that are pure dissipation in the sense that these activities consume resources while generating no identifiable increase in value anywhere else. Since no set of regulations can ever be comprehensive, individuals affected by the quota always tend to make resource-consuming adjustments. Even if a failsafe set of regulations could be devised, their enforcement would still not be absolute simply because this effort also consumes resources and at the margin there is always an incentive to cheat.

Thus far we have discussed the absolute size of the quota as if it were determined exogenously. Strictly, its actual level will never precisely match the stated level or, perhaps, even that intended level which the cartel would choose if faced only by the nonprice discrimination constraint. With the addition of enforcement costs, the reduction actually achieved will be less than would be indicated by simple analysis of the usual elasticities.

One method for participants to violate the regulations would be to overstate their positions with respect to the allocation criteria. With the total held constant, this method could only change the relative positions of quota owners, resulting in some reduction in the quota rent when this process proved resource-consuming. At the outset of the quota, this type of adjustment was apparently widespread, for the total level of imports was consistently above the stated intention of the Department of Interior. In fact, the official 12.2 percent level of imports (relative to domestic production) was arrived at only after it had been determined in a de facto manner—the original level nominally being 10 percent.

Enforcement costs are themselves a function of the quota size or, rather, of the value of cheating on the quota via any of the numerous alternative methods available. These alternatives encompass both marginal and total infractions of the stated regulations. Although it is difficult to distinguish between the two types of violation, we attempt this dichotomy because we believe that two separate types of resource expenditures are involved.

Of the numerous forms that violations of the quota can take, clearly the most costly to owners of the quota would be a lawsuit (or other action) that would abolish the entire program. Several actions were started, but all challenges in the courts were either withdrawn or moot. Viewed in tandem with the behavior of the OIAB and the OIA, this record suggests that a primary function of these bodies may have been to adapt the program in line with changing incentives so as to avoid abolishment or modification, preventing ownership of rights under the quota from becoming valueless.

The decision of the board on Atlantic Richfield's petition for additional quota illustrates both the difficulty of establishing comprehensive regulations and the hypothesis that the board's function was not regulation per se, but rather the enforcement of property rights. In response to Atlantic's request, which seems to have been entirely consistent with the quota regulations, the board said:

> Faced with the situation in which the OIA may not have consistently identified certain materials, the Board believes that its responsibility for review can best be served by adhering to the obvious purposes of the refinery and petrochemical phases of the program. It was clearly not the purpose of the Regulations to issue petrochemical allocations based on the production of gasolines. Therefore, it is the opinion of the Board that, both in terms of *the letter and the spirit* of the regulations, the disallowance of petitioner's claim respecting Plant 6 was not characterized by error.[5]

In other words, the presence of loopholes would not be allowed to jeopardize the rights of quota owners.

The OIA's compliance section was the official policing agent for the quota. It had help from both the Coast Guard and, particularly, from the U.S. Customs in barring the landing of unauthorized shipments of oil; but the internal group was ultimately responsible for assuring that the stated domestic throughput of the refiners and/or their historic importing records were accurate.

Standardized reporting forms and import licenses (to be inspected and certified by the Customs official) reduced the enormity of the task, but perhaps the most effective policing was accomplished by member-owners of the quota, since the OIA lacked both manpower and expertise to perform this task fully.

The controversy over definition of quota-earnings inputs for the petrochemical industry abundantly demonstrated the latter deficiency.[6] Each company applying for import rights filed a statement to justify its request. The number and the complexity of inputs that a refiner could claim as eligible created such inherent difficulties for the OIA that the enforcement even of reporting accuracy must have kept its staff of three full-time members busy.[7] Effective enforcement costs money; thus it is not surprising that $75,000 of supplemental appropriation was sought to augment the OIA's "surveillance and inspection program."[8]

Fortunately, the agency appears to have had at least occasional assistance from interested parties, as in the case where two companies lodged complaints

---

5. *Oil Import Digest,* Case No. Q-73, 1968, p. B-438 (italics added).
6. See following section, "The Petrochemical Challenge."
7. *Oil and Gas Journal,* May 1, 1967, p. 102.
8. Ibid., February 17, 1969, p. 47. This appropriation was not granted.

that the quota's provisions were being violated due to "confusion" in the reporting of natural gasoline. Other producers, they claimed, were reporting this as part of their domestic throughput in order to earn extra tickets. The accusations were checked by comparison with Bureau of Mines figures on stated input, and in 1966, the Oil Import administrator, Elmer L. Hoehn, addressed a letter to some 150 refiners in response to such complaints of systematic overstatement of their refinery runs.[9]

Without such industry surveillance there was evidently little threat of being caught in a violation. Not until 1968 did the enforcement section take any independent action. At that time Standard Oil Co. (Indiana) received the dubious distinction of being the first company audited for possible misstatement of quota-earning inputs.

Quota violations could bring penalties, but on the whole these were modest. Interior once asked for a 507,610 bbl. return in future quota. Importers normally ran the risk of losing a maximum of 1 percent or of 100 bbl. (whichever was smaller) of future quotas in the event they were caught in misleading, late, or false reports of inputs, or if they acted under exchange agreements before reporting to the OIA.[10] Although there is no record of the precise mechanism set up for enforcement of the division of the quota, it is clear that the OIA's direct participation was minimal.

Enforcement governing the total quantity of imports involved a somewhat different issue. Industry's concern in this process was clearly evident in its response to legal challenges against the quota, as illustrated earlier in the Texas-American Asphalt case. Still more conspicuous was its role during an attempt by New England's governors and the Consumers Union to have the quota dismantled.[11] In this case, the oil companies intervening in the suit on behalf of the federal government were aided by the oil-state governors of Louisiana, Oklahoma, Texas, and Wyoming.[12] The industry was represented by members of the Committee to Support the Mandatory Oil Import Program, through Baker and Botts, a Houston law firm. As in the Texas-American Asphalt case, no legal decision was handed down; this time the case was mooted when the Department of Interior allowed an increase in imports of distillate fuel oil.

The New England governors' suit was just one of many assaults on the quota. Earlier battles had raged over the control of finished products, the first being fought over residual fuel oil. By the mid-1960s, heating oil presented the major problem for the OIAB to resolve. In April of 1966, President Johnson

9. Ibid., November 7, 1966, p. 58.

10. Ibid., December 2, 1968, p. 48.

11. This suit, filed in U.S. District Court in Maine, contended that the quota practiced regional discrimination and thus was unconstitutional.

12. *Oil and Gas Journal,* December 1972, p. 45.

essentially decontrolled residual fuel oil, the lowest priced oil product on the eastern seaboard.[13] Companies lacking both an importing history and deep-water terminal inputs were made eligible, and individual allocations were increased. In addition, an extra reserve of 8725 b/d was granted to the OIAB for special appeals and, to further ease the restrictions, anyone wishing to import was granted the right even after the fact, if he presented the administration with proof of delivery.[14]

Interior Secretary Udall also announced an expansion of the OIAB's kitty to be used for new residual fuel quotas and the increase of existing ones. Newly eligible for quotas were consumers and governmental agencies in all districts.

Residual imports, meanwhile, had increased from 610,000 b/d in 1959 to some 950,000 b/d during 1965. Imports for the year ending March 31, 1966, were 800,000 b/d for District I alone. Domestic residual production, conversely, had been continually falling to the point where only 8 percent of an average barrel of crude was processed into residual oil (compared to 25 percent in 1946, for example).

Immediately following the relaxation of residual controls on the East Coast, the OIAB approved several additional import licenses, the first of these going to New Orleans Public Service Inc. for 350,000 bbls. on the grounds of exceptional hardship. This fuel was reportedly to be used for the utility's reserve in case of interruption in the supply of natural gas. These actions of the board, the president, and the Department of Interior, coincided with increases in the prices of residual on the eastern and gulf coasts and with threatened price increases by the Venezuelans.[15]

Apart from the maneuvers with residual fuel oil, little disturbed the operations of the quota during the early sixties. However, this comparative peace was shortly broken by two major controversies. By 1968, the import of residual fuel oil was still relatively unrestricted, but its lighter, more volatile and more expensive cousin, distillate fuel oil, was still under severe controls. Appeals began pouring in asking the board for additional distillate quota due, of course, to exceptional hardship. In August of 1968, a frequent source of additional tickets was once again tapped and 4000 b/d of extra product quota valued at close to $2 million a year were transferred from the Defense Department's allotment to the OIAB for allocation.[16]

---

13. Ibid., April 4, 1966, p. 103.
14. The actual requirement was more complex since, to be granted a license, an importer had to meet several qualifications. He must own the oil at the time of delivery to the terminal; the delivery must be the first one made of that oil into a deep-water terminal in District I; and he must also have delivered 25,000 bbl. of residual into the terminal under the agreement during the allocation period ending March 31, 1967.
15. *Oil and Gas Journal,* January 17, 1966, p. 29.
16. *Oil Import Digest,* National Petroleum Refiners Association, historical volume, Washington, D.C.

The other challenge to quota owners came from outside the refining group. About the same time that the distillate controversy was heating up, the producers of petrochemicals also began an attempt to reallocate rights to the quota. Both proved to be major attacks on the structure of the quota, and eventually the instigators of these actions were rewarded with compromise decisions that were nevertheless victories.

The early actions by Secretary Udall to release part of the distillate quota set the stage for the award of tickets to East Coast terminal operators whose applications were already before the OIAB.[17] As early as February 1968, twelve new licenses had been granted, totaling 6,973 b/d. As usual, the criterion was inability to obtain oil at fair (i.e., profitable) prices. Twenty-seven additional petitioners made the same plea, and the board responded by issuing a consolidated decision where additional tickets for light fuel oil were awarded, "owing to the unprecedented number of light fuel oil petitions which (the OIAB) has received this year." Northville Dock Corporation, which was included in this decision, was later to be cited as the precedent for numerous additional awards that followed.[18]

It seems that as the quota matured in years so did both the American Petroleum Institute (A.P.I.) gravity and the price per barrel of the product which was the center of controversy. By the late sixties, No. 2 fuel oil replaced residual as the major bone of contention; by the seventies, gasoline was the major concern.[19]

The changes that followed the early presidential proclamation on products applied for the most part to residual oil imports. This outcome should have been anticipated, prima facie, for several reasons: (1) fuel oil purchasers are fewer in number than other product users; (2) utilities, and large furnace and boiler operators, are the only significant users of residual oil because of the costs of extra equipment required to burn this fuel; (3) the area where fuel oil was, and still is, in greatest demand is the Northeast, where natural gas is most costly.[20]

Given the physical similarities between the various commodities restricted by the quota,[21] and given also the (apparently) greater incentive of residual oil users to violate the quota, it is not surprising that this was the first category to receive a variance from the original provisions. Lower costs of forming a joint

---

17. See Consolidated Interim Decision of the Petitions [of 18 companies for "immediate relief"], *Oil Import Digest,* February 12, 1968, p. B-363. Twelve companies were granted actual allocations, but additional cases such as United Oil Manufacturing Company, ibid., Case No. Q-3, p. B-367, and so on, should also be included in this move, since they were granted interim allocations on the same basis.

18. See ibid., September 27, 1967, Case No. P-15, p. B-341.

19. Evidenced, for instance, by the increased number of appeals for gasoline allotments.

20. Natural gas is a good substitute (technically) for fuel oil to most bulk users.

21. Implying similar transportation and smuggling costs, at least in one important dimension.

action between residual buyers (as opposed to gasoline buyers) would assure that the costs of enforcing the residual quota were correspondingly higher; thus (assuming that gains from the quota are comparable for all categories of oil) petitions from this area would predictably have been looked on with greater favor. A possibility, too, is that controls within the lower gravity categories might have been violated more profitably.

Our hypothesis of substitution within the regulated commodity—oil—implies that the quota will induce a greater percentage increase in the price for lower than for higher quality. To illustrate this, assume that five grades of oils are controlled by the quota, and that their prices per unit (i.e., per barrel) range from a low of 20 to a high of 100. The quota will increase the prices for all of these. Assume that, before any intercommodity substitution takes place, all prices increase 25 percent, making the new price range 25 for the lowest and 125 for the most expensive. For a given level of foreign prices, the import of the highest grade oil is most profitable.

When relatively more of the higher grade oil is brought into the United States, its relative price will decrease as a result of the quota.[22] Again for illustrative purposes, assume now that when all such adjustments are completed the lowest price will be 30 and the highest 110. Now, holding constant the foreign prices and freight, the most expensive oil will be only 10 percent higher than the foreign price, whereas the lowest quality oil will be 50 percent higher than the foreign price. This is illustrated in table 1.

TABLE 1

Effect of Quota on Relative Price

	Price before quota	Price with quota, not allowing intercommodity substitution	Price with quota, allowing substitution	Percentage increase above delivered foreign price
Highest grade	100	125	110	10.0
	80	100	90	12.5
	60	75	70	16.7
	40	50	50	25.0
Lowest grade	20	25	30	50.0

22. This follows because the quota is stated in barrels, and the equilibrium condition is a constant per-barrel differential between foreign and domestic prices prior to the quota (as well as subsequent to it). See chapter 6 for a more detailed explanation.

Individuals seeking to exploit the price differentials created by the quota could, at some cost, bring oil illegally into the country. In our example, a smuggler would gross $10 per unit on any of the grades; however, each dollar he invested in bringing in higher grade oil would yield (gross) only 10 percent, compared to a 50 percent return for the lowest grade. In our example, we held the relative prices between foreign grades constant; however, they will in fact move in the opposite direction to the domestic prices, further increasing the percentage differential.

Behind this adjustment process lie two necessary conditions. One is that some amount of each of the oils be imported both prior to and following the establishment of the quota.[23] The other is that the quota be stated in terms of rights to import barrels of oil (with *oil* defined in such way as to include all of the grades in the example). "Barrels" are the margin explicitly enforced under the quota, but a maximizing oil trader must ultimately be concerned with the dollar margin. The enforcement of the barrel margin will enter as a constraint into his decision process, but so long as differential rates of return exist for alternative grades of oil, resources will be spent disproportionately between grades. The equi-margin principle implies that these differential rates would cause relatively more resources to be spent to obtain tickets for the lowest grade oil, legally or otherwise. Though this incentive alone does not adequately explain residual oil's early release from the quota, it is a force working in that direction.

Were there no costs of enforcing the quota, the adjustments would end once the drift toward the high spectrum of oils was achieved. However, once we allow for the costs of authenticating records and authenticators, quota owners would never find it profitable to entirely eliminate violations, thus allowing adjustments in addition to the shift in oil quality.[24]

By smuggling we simply mean any method of violating the quota's provisions, and given diminishing returns to smuggling, some of this resource-consuming activity will occur at each of the various margins. The displacement of less preferred petroleum is not such an infraction, but this process succeeds in provoking the pattern of expenditures postulated.

Returning to our story, the reduced control of distillate fuel oil did not restore tranquility under the quota. The autumn of 1967 found northeastern senators and Hess Oil, together with several independent marketers, appealing to the OIAB and directly to Interior Secretary Udall for the granting of additional tickets for this product to be imported into the Northeast.

Leon Hess, chairman of Hess Oil and Chemical Corporation, offered to

23. This condition excludes perfect substitution between grades in the illustration.

24. We must ignore the question of why the quota would not be so tightly specified as to prohibit all adjustments if enforcement costs actually were zero. But if these costs are nonexistent, there probably would never have been a quota to analyze.

supply an additional 25,000 b/d of No. 2 (light heating) oil in return for these additional tickets. The company's Virgin Islands refinery would, reportedly, offer the additional oil at prices that would "never be higher than that at which major and independent oil companies are selling to their established customers." Hess went on to say that he would even "buy out of our present commitments [to European customers] to make this available."[25]

Excluding the Hess proposal, 1.5 million bbl. of finished products were being allocated to those ruled eligible by the OIAB for the remainder of the year, and Northeast Petroleum Corporation (Boston) was advancing proposals of its own, including an offer to import two million barrels of No. 2 oil.

While the Northeast group was pleading for additional tickets, independent producers were pushing a bill sponsored by Senator Russell B. Long (Louisiana) to commit to law the 12.2 percent ratio of imports to production for Districts I–IV. This bill would, at least nominally, have reduced the discretionary power of the president and provided the entire quota with an explicit legislative basis. It should, however, be noted that, since the total level of consumption is necessarily a prediction (subject to bias), it is not clear that the mere specification of the product of the total and this legally fixed percentage would have yielded a different level of imports.

Countering this congressional move was the Feldman-Kennedy group, submitting a proposed amendment to the Trade Expansion Act to partly decontrol No. 2 and to give the president or Secretary Udall discretionary power over the level of imports upon the filing of a report stating that "any likely price increases [resulting from the secretary's action] would not harm the economic welfare of the nation or national security."[26]

The *Oil and Gas Journal,* echoing oil interests, reacted with alarm.

> Demands by independent marketers on the East Coast are part of a pattern of raids on the control program. These marketers, with influential lawyers and politicians in their camp, sense a weakening of the system. They are striking while they consider the Interior Department to be in a vulnerable position . . .[27]
>
> Robert Kennedy's office told the Journal . . . that he expects to "win this one" with Interior, as he and others won the battle to lift contols on residual earlier. . . . The industry should know the Northeast group is playing for keeps. The Kennedys have somehow got the idea—heaven knows where—that the import program responds to political pressure. . . . They are after permanent access to foreign heating oil. . . . . At this point, don't bet against them. It is not at all certain that Interior and White House will be able to hold them off.[28]

25. *Oil and Gas Journal,* September 25, 1967, p.16

26. Ibid., September 25, 1967, p. 77.

27. Raids on the OIAB's allowable ticket allocation for finished products, such as the consolidated decision of September 27, 1967 *(Oil Import Digest,* p. B-349 of that date), had exhausted its quota.

28. "Watching Washington," in *Oil and Gas Journal,* November 13, 1967, p. 73.

By December, and immediately following the appearance of this summary of the foray, Secretary Udall had granted Hess a unique right to import 15,000 b/d of No. 2 during the winter months in exchange for a promise to invest some $70 million in refinery, petrochemical, and related facilities in the Virgin Islands.[29]

The controversy did not end here. The following year (1968) found the secretary and the OIAB printing additional tickets (some 4000 b/d) for eastern terminal operators, with the additional tickets coming, again, from the Defense allocation. Additional allocations and modifications of the import program were made over the following years to allow for increasingly larger imports of No. 2 oil into the Northeast.

These changes were of less import than the precedent-setting actions during 1967, but a few of them merit attention. In 1969, crude importers were granted extra tickets to be used in obtaining domestically produced No. 2.[30] In 1972, the Office of Emergency Preparedness suspended this restriction on the ticket use and simultaneously abandoned the provision that all imported No. 2 oil be produced in the western hemisphere.[31] But perhaps the major legal change came June 17, 1970, in Presidential Proclamation 3990, allowing independent terminal operators and independent marketers with access to deep-water terminals to import No. 2 oil. This action clearly redefined property rights in the quota.

The case of asphalt provided an exception to the usual relationship between oil gravity and import quotas over time. However, asphalt is both more expensive and more costly to transport than other heavy oils. Until 1967 there had been little controversy over the regulation of this relatively unimportant product. A small flurry of interest occurred when in 1967 the OIAB issued a 900 b/d quota to John J. Hudson after having dismissed two earlier appeals by the company.[32] As usual, exceptional hardship was cited as the cause, but this reconsideration was significantly timed, since one month earlier Secretary Udall, following President Johnson's authorization, had announced plans to boost asphalt imports. It led some (at least the *Oil and Gas Jounal*) to suspect that a general relaxing of restrictions was to follow.

However, little excitement was generated in the industry journals even when in September 1969 major additions were made in the asphalt control quotas, with the bulk of new tickets going to District I, which again became

29. Ibid., December 18, 1967, p. 60. This is an oversimplification of the terms, and later Hess was to be "caught" violating the agreement.

30. Ibid., January 27, 1969, p. 93.

31. No. 2 oil ticket holders could no longer exchange foreign crude for domestic distillate; by 1972, these tickets had again become applicable to the importation of distillate oils only.

32. *Oil Import Digest*, May 5, 1967, p. B-345. John J. Hudson Co. had filed suit in U.S. District Court for the District of Columbia, challenging the decision of the board and seeking the right to import finished product asphalt after being denied by OIAB. Ibid., Case No. 0-30, December 23, 1966, p. A-181.

the geographic center of controversy.[33] Prior to 1970, very few appeals had been presented for additional asphalt tickets, but by July 20 of that year, under a barrage of sixteen appeals, the board had awarded a total of 2,555,500 bbls. of finished asphalt, all on the familiar hardship grounds.[34] This increase more than doubled the original allotment of asphalt tickets.

Gasoline never seemed to provoke the interest that fuel oils commanded during the sixties. Appeals were presented to the board for gasoline import rights over the entire period, but apart from the Puerto Rican arrangements, few major awards were made[35] until 1972.[36] By February of 1973, the board was described as "spewing" gasoline quotas.[37]

In 1966, the self-policing and adaptability issues were illustrated once again. Commonwealth and Gulf Oil corporations began importing an extra 10,000 b/d of gasoline from Puerto Rico into District V by exploiting a loophole in the quota regulations. This action quite understandably prompted a stream of complaints from other District V importers whose quotas were thereby reduced on a barrel-for-barrel basis. But the controversy ended quickly when Secretary Udall, Commonwealth, and Gulf agreed to transfer these shipments to Districts I–IV while holding them at the already effective level of 10,000 b/d.[38]

THE PETROCHEMICAL CHALLENGE

Until around 1964, the question of petrochemicals scarcely intruded in discussions surrounding the oil import quota. Worldwide, the petrochemical industry is based mainly on two alternative petroleum-based raw materials—natural gas or naphtha, the latter a product of crude oil. For many years, and independently of the quota, the United States petrochemical industry had exclusively used the less expensive natural gas.

In the early sixties natural gas became less popular, due partly to technological changes and partly to a shift in relative prices. Since the import of naphtha fell under oil quota regulation, its domestic price rose substantially above the international level, thus offering the petrochemical industry a new and "valid" reason to claim a share of the quota.[39]

Standing alone, such a reason did not mean much; every consumer of oil

33. John J. Hudson Co. was granted 514,300 bbls. of finished asphalt under Section 3(d) of Presidential Proclamation 3279, as amended to allow the Secretary of Interior to allocate asphalt without licenses required. Ibid., Case No. S-1, April 6, 1970, p. B-520.

34. See Ibid., Consolidated Decision, July 20, 1970, p. B-647-59.

35. See following section on petrochemicals.

36. *Oil and Gas Journal,* September 11, 1972, p. 71.

37. Ibid., February 15, 1973, p. 26; and May 16, 1966.

38. Shaffer, *Oil Import Program,*

39. By 1962, Union Carbide apparently already had applied for, and had been denied, a naphtha quota. *Oil and Gas Journal,* December 21, 1964, p. 42.

products was suffering similarly. However, the U.S. petrochemical industry, as a major exporter, could argue that foreign consumers faced with higher prices for their products might easily switch to local sources of supply. Moreover, since the domestic market for petrochemicals was not heavily protected, imports might replace some local production. If, as it appears, the demand for domestic naphtha was highly elastic, binding the petrochemical industry within the restrictions of the quota would in any case have yielded only low returns to the oil industry.

Union Carbide's attempt to take advantage of free trade zones (mentioned earlier) had been a preliminary effort to escape some effects of the quota. It is worthwhile citing at some length a report of the event:

> A plan to extend the foreign trade zone at New Orleans to a point some 20 miles up the Mississippi River has prompted Union Carbide Corp. to select Taft, La., as the site of a $50 million petrochemical plant.
>
> Union Carbide expects, among other things, to import naphtha without restriction from Venezuela into the Taft plant. The company then will convert the naphtha to ethylene for unrestricted marketing in the U.S. as well as for export.
>
> Ordinarily naphtha is subject to import controls—but not into a trade zone. There are no controls on ethylene.
>
> Union Carbide, which has a similar operation in Puerto Rico, apparently is the first company preparing to take advantage of a domestic trade zone for petrochemical production, and its plan has touched off a flurry of interest among domestic petroleum and petrochemical industries.
>
> Some may follow suit with plants of their own; some may oppose the Union Carbide program on the grounds that it puts competitors at a disadvantage.
>
> *New Legislation.* Extension of the trade zone to Taft is made possible by a law just passed by the Louisiana legislature. The law authorizes the Port of New Orleans to extend its boundaries upriver into adjoining St. Charles Parish where Taft is located.
>
> Once this is done, the port authority would then apply to the Foreign Trade Zone Board, a federal agency, for extension of the foreign trade zone area to cover the new territory.
>
> Union Carbide is aware that a number of federal agencies, such as the Oil Imports Administration, will be checking the validity of the company's import, processing, and marketing program. Some of the ramifications of the program are still unknown.
>
> However, most authorities believe there are no restrictions on the import of chemicals. And if Union Carbide processes naphtha, a hydrocarbon, into

> ethylene, a chemical, it can then move the ethylene into U.S. markets without restrictions.

And a few paragraphs farther:

> *Flurry of interest.* Other petrochemical companies with plants along the Gulf Coast, other U.S. port areas, or in Puerto Rico are doing some fast checking to determine whether they should fight such a move or get in on the act themselves.
>
> Federal officials say they are getting calls, visits, and queries from a number of firms, including some "very big ones."
>
> Oil firms with petrochemical facilities fear a possible surge in competition from petrochemical plants that have no oil-import quotas and thus have been foreclosed to using foreign oil as a raw material.
>
> Such plants using trade zones could bring in controlled materials, such as naphtha, without quotas and use them in producing nonquota petrochemicals for sale within the U.S.[40]

Virtually every facet of the described activities represent the use of resources in an attempt to escape some effects of the quota. From among the potential foreign-trade zones Union Carbide chose Taft, Louisiana; whether that was the best choice remains an open question. In any case, both Union Carbide and government agencies had to spend resources to convert the area into a foreign-trade zone. The need to clarify uncertainties and to check the validity of the company's program required some resource expenditure. So, also did the studies by other firms "whether they should fight such a move or get in the act themselves" and the matching research undertaken by government officials.

What was being reported here was clearly only the opening shot of a long fight. In that struggle the departments of Commerce and State aligned with the petrochemical industry, whereas Interior gave its support to the oil industry.

Various hearings were held. The first, on December 14–15, 1964, by the Commerce Department, dealt with the confirmation of Taft as a foreign-trade zone.[41] A similar hearing on June 8 considered a request by Dow Chemicals for a similar zone in Michigan.[42]

On October 28, 1965, the Deartment of Interior heard some thirty-two witnesses,[43] most of whom were highly placed executives of major petro-

40. Ibid., July 20, 1964, p. 52.
41. Ibid., December 21, 1964, pp. 39–42.
42. Ibid., June 14, 1965, pp. 80–81.
43. Ibid., November 8, 1965, pp. 82–83.

chemical and oil firms.[44] The secretaries of Interior and Commerce, as well as many of their subordinates, were present. The end result was a decision to provide petrochemical companies a share of the quota, while also imposing a quota on imports of oil into foreign-trade zones.

The Union Carbide decision did not end the foreign-trade-zone controversy. In fact, the most hotly debated application was not to be resolved until the early 1970s, after Occidental Petroleum Corporation had applied for the right to establish a 300,000 b/d refinery in Miachiasport, Maine. This request precipitated a struggle that brought virtually all of the major oil groups into the limelight. Along with the previously mentioned groups such as IPAA, Tipro, and the National Petroleum Refiners Association (NPRA), were the API and several individual companies, all opposing Occidental's plan. On the other side were the New England governors (again), Occidental, and several miscellaneous political figures. Occidental never built the refinery, and the flood of similar applications welling up around the nation dried out following Nixon's election.[45]

The quota allotted to the petrochemical industry came at the expense of refiners—the overall quota remained intact. The petrochemical quota resembled the refinery quota in two major respects: it was based on inputs, and initially it amounted to approximately 10 percent of these. It differed from the refinery quota, however, in lacking any sliding-scale provision. More importantly, it presented the insuperably difficult task of pinning down exactly what petroleum inputs enter into a given petrochemical, and of measuring them. This complexity gave ammunition to the petrochemical industry in its continuing battle to obtain 100 percent quotas, which it considered much more satisfactory than any allotted 10 percent (however defined).

The first headaches appeared in trying to establish a base for petrochemical quotas. Any chemical firm uses a large array of inputs, one subset of them almost certainly petroleum based. In any refinery, crude petroleum obviously dominates all other inputs; but in other chemical plants, including petrochemicals, the share of petroleum is much smaller. If quotas were to be assigned to every firm chemically converting petroleum products, the number of qualifiers might be immense and their regulation prohibitively expensive.[46] Because the petrochemical lobby was apparently dominated by the largest

---

44. For instance, Union Carbide sent its executive vice president, and Atlantic Refining, Du Pont, and Conoco, their vice presidents. Ibid., November 1, 1965, pp. 30 – 31.

45. Ibid., June 14, 1966; July 29, 1968, p. 42; and September 26, 1966. Also, see Dam, "Implementation," pp. 47–48.

46. While combustion is also a chemical reaction, it is usually, and seemingly fairly easily, separated from other chemical reactions.

firms, the formula finally reached would be expected to favor that group.[47]

On December 10, 1965, the Presidential Proclamation was amended to accommodate petrochemicals, effective January 1, 1966. The method of allocation appears deceptively straightforward. Petrochemical plants were to receive quotas "amounting to about 9% of their inputs of condensate, ethane, propane, butane and naphtha."[48] Given this simplicity, one would have expected individual company quotas to be announced by the effective date. Apparently, the actual translation proved so difficult that two months later, in March 1966, the Department of Interior drastically modified the base for qualification.[49]

It is difficult even to grasp the meaning of the change in regulations; this is partly due to the spate of polysyllabic chemical terms, which in themselves underscore the complexity of the case. The petrochemical industry obtains some of its raw materials from crude oil derivatives and some from natural gas. Some refiners process petrochemical feedstocks in their own plants. What, precisely, is to be included among qualifying inputs? How large should they be for the plant as a whole to qualify?

It took the Department of Interior until September 1966 to come up with the actual allocations, retroactive to January 1. Even this long gestation period did not produce a definitive solution. For instance, at the end of 1966, the definition of a petrochemical plant was expanded to allow an output measure as an alternate to the previously established input measure. In July 1968, the output of some plants used as inputs in others was excluded from the quota base. Such instances are abundant.

Let us turn now to what was in part a quirk in the program—but nevertheless a significant one in terms of the way it was handled—the allocation to Standard Oil of Indiana Petrochemical. Standard, in its refining activity, operated catalytic reformers. In these reformers, naphtha was used as an input, and more than 50 percent of the output comprised benzenes, toluenes, and xylenes, variously known as BTX, which qualified for the petrochemical quota. These in turn were used (in lieu of lead) to raise gasoline octane rating. Standard qualified for a quota first as a refiner and then as a petrochemical plant—a double dip into the overall quota pool. On the basis of the production of BTX it was given a quota of 6708 b/d for 1968 (as well as retroactive quotas of 3458 b/d for 1966 and 2853 b/d for 1967). At $1.25 per

---

47. This conclusion is reached by observing who financed the lobbying efforts. Arthur D. Little, Inc. was first commissioned by Dow Chemicals and Monsanto to prepare a study of petrochemical feedstock, and subsequently the same consulting firm was commissioned by Dow and by Union Carbide to rebut a study for refiners. *Oil and Gas Journal,* August 22, 1965, p. 56.

48. Ibid., December 20, 1965, p. 40.

49. For a discussion of this modification, see ibid., March 28, 1965, p. 105.

barrel, the commonly cited value of a ticket, the value of the quota for 1968 was some $3 million.[50]

This double quota was denounced by Senator Proxmire on February 28, 1968, and within five days, Standard's petrochemical license was revoked by Interior Secretary Udall on the basis that the application had been filed late. This brought a counter action from Standard in the form of a lawsuit, whereupon the Department of Interior withdrew from its earlier position.[51] However, Standard subsequently lost the fight when the regulations themselves were changed to prohibit the so-called double dip.[52]

Even while efforts to achieve a workable set of regulations were in progress, oil refiners and petrochemical manufacturers were engaged in unfraternal struggles to increase their respective shares of the quota. In an attempt to demonstrate, among other things, that they could not compete in world markets unless free access were granted to foreign raw materials, the petrochemical producers marshaled Arthur D. Little, Inc., to study and present their cause. Thereupon the oil industry commissioned its own study which was completed in July 1967. One of its conculsions was: "If U.S. petrochemicals manufacturers had unlimited access to foreign naphtha, most of them would not be interested in purchasing it at present price levels."[53] It is anticlimactic to observe that the oil industry still demanded restrictions on the import of naphtha.

In all, the petrochemical industry sponsored at least four studies by Little.[54] The oil industry, in addition to the study just mentioned, commissioned Stanford Research Institute for another, completed in August 1967.[55] Between times, individual firms as well as trade organizations in the two industries were busily preparing their own reports, many of which were presented in various hearings or were submitted to commissions.

An instructive turn of events occurred late in 1971. We have seen that in the sixties, as the industry expanded, naphtha had become a more competitive raw material, largely replacing natural gas in new plants. By 1971, the price of (unregulated) natural gas had increased until it suddenly became profitable to convert naphtha into gas (more specifically, into SNG, or substitute natural

50. See ibid., March 4, 1968, p. 64.

51. Ibid., March 11, 1968, p. 64; April 18, 1968, p. 46; and May 27, 1968, p. 53.

52. Had the BTX operation been under separate corporate control, a separate quota on the oil-based inputs presumably would have obtained and the double dip issue would never have been raised.

53. Cited in *Oil and Gas Journal,* July 11, 1965, p. 66, from a study by Petroleum Industry Research Foundation, Inc. (PIRINC), which was funded by East Coast Terminal Operators and some major oil companies.

54. The dates for these studies are: August 1965, August 1966, and (two studies) June 1969. See A. M. Kirby, Jr., "The Chemical Industry and the Oil Import Program."

55. Some details of the study, completed in August 1967, are given in Kirby, "The Chemical Industry."

gas). Obviously, such conversion was more profitable if imported naphtha could be used.

Was such use of naphtha covered by the original proclamation or by any of its amendments? Not explicitly, for this development had never been envisioned by the regulators. Whether any of the more general rules could be construed as applicable seemed subject to enough doubt so that several applications were submitted for the importation of naphtha to be converted to gas. Some, but not all, of these were submitted to the Oil Import Apppeals Board; others were submitted to other agencies, presumably because the applicant felt that jurisdiction (maybe practical rather than legal) lay elsewhere.

TWO AD HOC EXAMPLES

Puerto Rico. Its position within the Customs territory of the United States placed Puerto Rico under control of the mandatory program. However, at the outset, a special status was granted to Commonwealth Oil, Puerto Rico's major refiner, because of its total traditional dependence upon Venezuelan crude oil. By 1965, Phillips Petroleum was allowed to join this privileged category although theirs was a much different arrangement.[56] Subsequently, Sun Oil in 1968 also achieved a special, but still different, status in Puerto Rico.

Both Phillips and Sun built petrochemical facilities designed to use western hemisphere feedstocks, with the proviso that their shipments to the United States be limited (in kind and quantity) to certain products. In addition, each of these companies was to invest in the construction of various facilities that would be surrendered to the Puerto Ricans following their completion.

Commonwealth and Phillips both exported gasoline to the United States from their plants, while Sun was allowed to ship products exclusive of gasoline. Union Carbide was granted a quota allowing unlimited petrochemical exports to the United States, but was explicitly prohibited from shipping gasoline to the domestic market.[57]

It seems that these exceptions to the quota were of value, since other companies also sought to participate in this scheme. Standard Oil Co. (N.J.), for example, cited the Phillips precedent in petitioning Secretary Udall for the right to help out with Puerto Rican employment problems.[58] That Standard's case did not fall upon sympathetic ears seems the only explanation for the rejection.

---

56. See Dam, "Implementation," pp. 44–46; and *Oil and Gas Journal,* March 29, 1967; December 18, 1967, p. 56; and June 28, 1965, p. 53.

57. *Oil and Gas Journal,* December 18, 1967, p. 56; this quota was accompanied by a promise from Carbide to expand its investment in Puerto Rico by some $50 million.

58. "Jersey starts 'me-too' line on naphtha for Puerto Rico," ibid., May 10, 1965.

Each case appears to offer a special, or ad hoc, example of Interior's discretionary power. The difficulty of formulating regulations that would cover all situations including the off-the-cuff reactions of the regulators when any deficiency was discovered, is illustrated by the Commonwealth episode.

Virgin Islands. Following the approval of the Hess proposal to import fuel oil into the Northeast, construction of a refinery in the Virgin Islands was completed in record time. The availability of lower-price Venezuelan crude oil as the feedstock may partially explain the haste, especially since the Jones Act did not apply to shipments from Latin America to the Virgin Islands.[59]

Apparently to avoid a deluge of petitions for similar privileges, Secretary Udall announced a closed-door policy regarding future Virgin Island quotas.[60] Hess in turn agreed to contribute fifty cents per barrel of the 15,000 b/d quota to a "conservation fund for beautification and pollution control."[61]

This modification to the quota came at the expense of all other refiners, since the 12.2 percent ratio of imports to domestic consumption was to be maintained. Accordingly, the IPAA denounced the Udall decision as bowing to special interests.[62] His action, the IPAA president said, provided "all the evidence needed by the Congress that pressures of individuals seeking privileged treatment under the import program will not be resisted in the absence of . . . legislation." The particular legislation referred to was the bill, as earlier briefly discussed, to commit to law the 12.2 percent formula. President Johnson, according to the *Oil and Gas Journal,* had earlier threatened to veto such a bill.

59. The Jones Act requires coastal shipping to be in U.S. flag vessels for which the rates are typically higher than for foreign vessels. (46 U.S.C. 861 *et seq.,* 1964).

60. *Oil and Gas Journal,* November 13, 1967, p. 71.

61. Ibid., November 20, 1967, p. 142.

62. Ibid., November 13, 1967, p. 70.

5 • The Allocation of Crude Oil: Oil Spoils

"It was a beautiful theory the authors of the import-control plan evolved—this idea of giving every refiner an import quota based on a percentage of his crude runs. . . . It was a share-the-wealth program that would make the whole industry healthy."[1]

Whether an individual was entitled to a share, and of what size, depended on whether he operated east or west of the Rockies, whether he was a historical importer, and whether he was refining domestic crude in the current period, and how much.[2] His share would vary from year to year and, for a refiner, would depend on how his particular size-class was being treated in the given period. Beginning in the mid-sixties, if the individual did not qualify under any of these, he could assert a variety of other qualifying criteria: that he required oil inputs into his petrochemical operations (and later, for his petrochemical exports), that he produced low-sulphur oil in the West, or simply that he was willing to operate in the Caribbeans and therefore was deemed to be deserving.

REFINING AND HISTORICAL ALLOCATIONS

These two classes within the allocation criteria differed widely in one respect: a recipient who wanted to change his share could legally expend resources to qualify for a larger quota under the refining criterion, whereas under the relatively simple one based on precedent, his share was parametric. Once established, the historical qualification rested squarely on the past record of imports in the middle of the voluntary quota period, 1957. The allocation to refiners, on the other hand, was so based that the operator who chose to increase his use of domestic oil could earn additional rights to import. Thus real resources could be expended merely to acquire property rights over the quota, but since the total level of imports was basically fixed, such expenditures produced no gains in the form of increased output.

The notion that the historical base for the quota is nondissipating is not

1. Editorial, *Oil and Gas Journal,* April 17, 1959.

2. A refining firm had the option of presenting as a basis for qualification either its domestic crude input or its historical imports.

strictly correct, however, even *ex post*. Resources might be expended over time to influence the choice of the base period, the definition of importer, and so forth. More importantly, it seems that the establishment of such a criterion would have been at least partly anticipated and that firms would have spent resources to qualify by increasing their imports above the normal range in 1956. This would lead to dissipation because importers who raised their imports beyond their previously optimal level were not necessarily the most efficient users. Since the historical criterion was widely adopted,[3] such anticipatory actions were probably common despite governmental warnings that in the inauguration of some new policy the recently chosen levels of operation would not be subsequently recognized as a historical base.[4] Despite these considerations, casual reasoning suggests that the historical base dissipates fewer resources than the refinery base.

Therefore, in an attempt to make the quota efficient, the government would have been expected to employ the historical base of allocation as the only, or at least the dominant, base. *Efficiency,* as conventionally defined, would dictate no quota at all: its inception indicates that other forces dominated. By the same token, we cannot really predict that, within the quota system, efficiency would be the only criterion, unless we can fully specify the underlying constraints. In fact, both the refining of domestic crude and the refinery quota itself expanded at the same time that the absolute amounts allocated on the historical basis were shrinking until, in 1971, the historical quota was abolished altogether.

Perhaps the explanation of this choice is related to the political power of small refiners. Within the refinery quota, smaller operators were relatively favored, whereas the historical quota served mostly to benefit large refiners.[5] The first observation implies that the smaller refiners had political power to effect transfers toward themselves. As additional evidence, consider the following.

When the quota was first imposed, those refiners operating in District I–IV (basically east of the Rockies) and refining less than 10,000 b/d of domestic crude oil received a quota equal to 12 percent of their input, compared with a marginal rate of 7 percent for refiners refining 100,000–150,000 b/d. By 1973,

3. For instance, when in 1970, Canadian imports east of the Rockies were subjected to quota, it was based on Canadian imports in 1969. Similarly on May 1, 1973, quantity restrictions were removed. Import fees were imposed in their stead, but not only were previous quota holders exempt—"historical" importers of newly covered products were exempt as well. Many other examples can be cited.

4. Skepticism seems justified. In the mid-50s, under the first voluntary quota, many firms quickly began to import oil; when the compulsory quota was established, these turned out to be counted as historical importers.

5. In 1954, Atlantic, Gulf, Sinclair, Mobil, SoCal, Standard of N. J., and Texaco, the established importers, imported about 90 percent of all crude oil brought into Districts I–IV. Their allocation for late 1957 during the voluntary program was about two-thirds of the total. U.S. Cabinet Task Force on Oil Import Control, *The Oil Import Question,* p. 190.

the rate for the small refiners had increased to 21.7 percent while that for the big ones had fallen to 3.8 percent. The time trends for the two rates were not entirely smooth, but the steady increase in the first and decline in the second are unmistakable.[6] The evidence is consistent with both the initial political power of small refiners and its increase over time.

THE EFFECT OF THE REFINERS' ALLOCATION

Historical quotas quite obviously were allocated to firms that had shown a comparative advantage in importing. If anything, the reverse was the case with the refiners, who qualified for quota strictly according to their input of domestic crude oil. Those using only domestic oil received a full share; those refining only imported oil received none. There was some debate whether to restrict refiners to use their quotas in their own refineries, or to make the quotas fully transferable. The actual decision lay somewhere in between. Transfers could take place, but with an element of barter—oil had to be traded for oil.[7] The cost of affecting the barter plays a major role in the explanation of some curious behavior to which we will come shortly.

To describe the nature of the exchange, in brief: an inland refiner holding a quota of one thousand barrels could buy oil from Venezuela and then exchange it with a coastal refiner in return for, say, 1250 barrels of Oklahoma crude.[8] Obviously, each of the parties had an incentive to obtain the oil suitable for the other's purposes. Yet, to demonstrate to the OIA that they actually exchanged oil, they had to record purchase contracts.

To the extent that such an exchange incurs a cost independent of its size, the exchange was more costly to small refiners. We have no direct knowledge of what this expense might have been, but clearly it was not trivial. A small domestic refiner, refining 2000 b/d, typically would earn a quota of some 15 percent of his oil input. The total semiannual quota, to be used within a six-month period, was then 15 percent $\times 2000 \times 180 = 54{,}000$ barrels. In absolute terms, this is a petty volume that could be carried in half the capacity of an uneconomically small WWII tanker.

Consider the possibility that this small refiner were to merge with a bigger

---

6. A similar pattern took place in District V. The rate for the smaller refiners increased, also quite steadily, from an initial level of 25 percent to 67.5 percent in 1973. For the larger-size class the rate was first 8 percent, rose to a peak of 17.08 percent late in 1964, declined to 1.9 percent in 1969, and then turned up to 5.6 percent in 1973. See Appendix A-l for the entire progression of the "sliding scale" in the two regions.

7. Since no money changed hands, the subsidization of refiners was partly concealed. This is considered by many to be politically advantageous. On the other hand, the figure of $1.25 as the per-barrel value of the quota was cited constantly.

8. Assuming that the two crudes are similar, and abstracting from transportation differentials, the difference of 250 barrels represents the net value of the quota to the inland refiner. Note that only ownership over oil changes hands.

one refining, say, 25,000 b/d. The marginal quota rate for the latter was, until 1964, about 2 percent less than that for the former. The loss of tickets due to merger for a six-month period would amount to 7200 barrels of oil valued at some $9000. Against this loss could be weighed the saving in transaction cost of now handling one exchange rather than two.

While the sliding scale in refiners' quotas provides an incentive to reduced size, the exchange requirement seems to favor the larger firms, and greater transaction costs might actually nullify the benefits accruing to a smaller firm from its higher quota rate. A merger with a refiner having East Coast capacity might make the exchange entirely unnecessary since the quota earned by the inland refiner could now be used by the same refiner in a coastal facility.

TABLE 2

Distribution of Refining Capacity by Refiner Size in Selected Years

Refiner size	**Districts I–IV**			
(thousand barrels/day)	**1957**	**1959**	**1967**	**1972**
0–30	15.94	15.77	11.19	8.68
30–100	11.11	13.45	11.94	9.47
100+	72.95	70.78	76.87	81.85

SOURCE: The following issues of *Oil and Gas Journal* containing "Survey of Operating Refineries in the U.S.": March 25, 1957; March 30, 1959; April 3, 1967: and March 27, 1972.

The data on refiners in Districts I–IV yield no clear pattern with respect to change in refiner size induced by the quota. Our attempt at such identification took two forms. First, changes in the size structure of refiners after 1959 were compared to those of preceding years (because of data problems, these are restricted to the period of 1957–59). The resultant pattern (shown in table 2) suggests that either the quota had no effect on the size structure or that, due to it, the relative position of big refiners improved. Between 1957 and 1959, the period immediately prior to the quota, the share of the largest refiners (over 100,000 b/d) had declined while those of smaller classes increased. The reverse held true for the two subperiods of 1959–67 and 1967–72.[9]

9. The choice of years is a bit arbitrary. The guiding principle is that by 1967, the differentials within the sliding scale were stabilized. The choice of size classes is the same that was in effect after the middle of 1962 (when consolidation of classes took place).

In a second attempt at identification, we utilized the sliding scale of the quota in terms of size of inputs of refining firms, using data on size distribution of both firms and plants. As the latter are not directly affected by the differential quota, we would expect their size trends to be less disrupted by its impact. The data on plant size (summarized in table 3) show no clear pattern; thus we conclude that this evidence for Districts I–IV is also consistent with the notion that the two forces affecting refiners' size more or less canceled each other.

TABLE 3

Distribution of Refining Capacity by Plant Size in Selected Years

Plant Size	**Districts I–IV**			
(thousand barrels/day)	**1957**	**1959**	**1967**	**1972**
0–30	20.90	18.88	12.17	9.07
30–100	35.62	36.98	38.16	30.74
100+	43.48	44.14	49.67	60.19

SOURCE: See source note to table 2.

District V differs from Districts I–IV in three important and relevant respects. First, the value of a ticket in District V seems to have been only about eighty cents as compared with $1.25 in Districts I–IV.[10] Given the sliding scale, the inducement toward the size reduction operated in both areas, but with less force in the West because of the smaller price differential there. Second, as indicated, in District V the sliding scale favored small refiners more strongly than in the East. Third, because most of the refiners in District V operated on the coast, the need to exchange oil was much less pressing.[11] This last factor means that forces countering the tendency toward smaller firms were largely absent in District V and, therefore, the presence of the sliding scale favoring smaller firms should dominate.

Tables 4 and 5 reflect conditions in District V and are used for performing the same kinds of test just described with respect to tables 2 and 3 for Districts I–IV. The distincton between firms and plants does not yield clear results. The test for change in the size distributions of both refiners and refineries, on the

10. Both numbers are by no means well established but, rather, are alluded to in various publications—much more frequently for the latter than for the former.

11. This probably explains why the value of a District V ticket is mentioned so infrequently.

other hand, produces rather strong results. The 1957–59 trend toward larger sizes is reversed, both for 1959–67 and for 1967–72. So, while no clear prediction could be made with respect to District I–IV, such a prediction is justified for District V, particularly as compared with Districts I–IV.

TABLE 4

Distribution of Refining Capacity by Refiner Size in Selected Years

Refiner size	District V			
(thousand barrels/day)	1957	1959	1967	1972
0–30	13.97	8.65	11.87	14.11
30–100	4.54	2.20	2.62	8.37
100+	81.49	89.15	85.51	77.52

SOURCE: See source note on table 2.

TABLE 5

Distribution of Refining Capacity by Plant Size in Selected Years

Plant size	District V			
(thousand barrels/day)	1957	1959	1967	1972
0–30	17.94	12.70	17.06	16.65
30–100	31.88	27.95	24.76	40.70
100+	50.19	59.35	58.19	32.36

SOURCE: See source note on table 2.

It was asserted earlier that, since the unit costs of exchanging oil decline with the number of barrels exchanged, this would exert a force counter to that of the sliding scale. The results on size distribution of refiners in Districts I–IV, while consistent with that assertion, certainly do not give it strong support.

Indeed, the hypothesis was derived *ex post* after observing the inconsistency of the results for District I–IV, with the predicted effect of the sliding scale. Consequently, it is desirable to test additional implications derived from the oil exchange hypothesis.

A ticket allowed its owner to legally import one barrel of foreign oil, provided that the right be exercised within the given time period, that permission be obtained prior to its exchange, and so on. To determine the value of a ticket, it seems that once we know the total allocation of tickets, Q_O, and the location of the demand function for oil, the equilibrium price is fully determined under competition and a zero cost of making these transactions. Figure 4 demonstrates this, where MV_t is the marginal-value schedule for tickets, Q_O is the total quota, and t_O is the implied value of a ticket. This marginal value schedule is an excess-demand schedule, in the sense that, as larger allocations of tickets are made, the foreign and domestic prices tend to equalize.

FIGURE 4

Equilibrium Price of Ticket under Competition

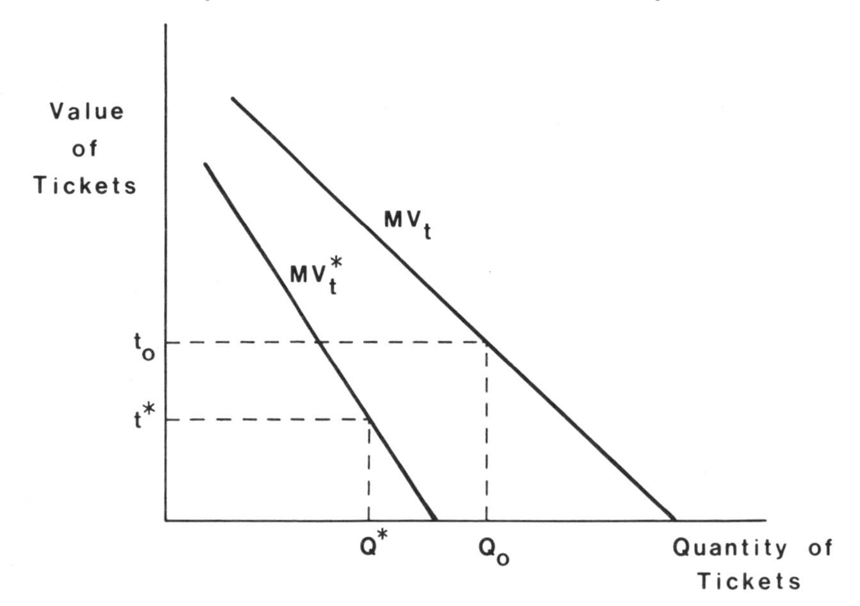

The positive ticket value depends on the price difference between domestic and foreign oils. By manipulating this difference, a testable implication follows. If freight costs were to increase causing MV_t to fall to, say, MV_t^*, the value of a ticket would fall from t_O to zero. Ticket demanders are not constrained to purchase the foreign oil, so the maximum amount that would be offered for these rights would be the difference between the delivered foreign cost and its domestic counterpart. At these new (higher) freight rates, ticket holders would find themselves holding the barrel, and unless the total allocation of rights were reduced, they would be of no value.

We wish to now turn to the question of whether all tickets will always be used. The maximization hypothesis implies that ticket traders could not find themselves anywhere but at the point where all gains from exchange have been exhausted. This equilibrium condition implies that at any positive ticket price the entire allocation will be used, for any unused ticket is simply wasted. There is no way within this framework that both a positive ticket price and unused tickets could exist simultaneously. In other words, given the allocation of Q_O and the original market-value schedule (prior to the freight rate increase) all points such as (Q^*, t^*) are logically excluded because holders of tickets would find it profitable to offer the additional Q_O - Q^* tickets for sale, as ticket buyers would willingly offer prices higher than t^* to encourage this. Thus for t^* to be the new ticket price, the total allocation would have to be reduced to Q^*, which is a variable beyond the control of ticket owners.

The reasoning behind this result is, of course, simply the basic theorem of exchange. We only emphasize it here to introduce an observation that should puzzle economists. Explaining it may add to our understanding of quota's effects (and perhaps our view of economics). To underscore the puzzle, imagine a similar situation, but instead of tickets, suppose that one-pound bars of gold were offered for sale at $1000 a bar at the same moment that other gold bars were free for the taking, yet the $1000 was still paid for some. Such an event would obviously be excluded by the previous reasoning, and if actually observed, would refute the theorem of exchange as it is used here.

To most inland refiners, the overland transportation costs canceled the value of their tickets if the foreign oil had to be used directly in their refineries. But inland refiners were allocated tickets, not oil. If they could sell their tickets, and this action entailed no cost, the entire rent from the quota, as just discussed, would be captured by ticket owners.[12] If such exchange is costly, some of the rent that previously appeared as gain to quota owners will be spent by consuming the foreign oil in less valued uses, by extra transportation charges, as extra searching by new importers for foreign oil sellers, or possibly as unused tickets. All this because refiners were granted the right to exchange but simultaneously were restricted to this curious barter of oils.

A brief excursion through the economist's classic prisoner-of-war camp may serve to illustrate the conceptual problem facing traders under the quota restrictions. This time, rather than exchanging cigarettes for candy bars to illustrate the theorem of exchange, our prisoners trade their cigarettes for

12. This may be viewed as a quasi-rent since here we ignore any resources spent to obtain the ticket. We also abstract from the domestic producer's rent.

cigarettes. If the resources required to undertake this activity are consumed solely from the cigarettes themselves, then cigarettes will become progressively shorter as the exchange proceeds. The rate at which the tobacco disappears measures the costs of transacting. Similarly, the quota specifically ruled out the sale of tickets in favor of a barter arrangement. Oil had to be exchanged for oil; thus an inland domestic refiner could only gain from the ownership of import rights by bartering for domestic oil. Whatever the ostensible reasons for installing this system, its effect was to add extra costs to the process of exchange.

Suppose that an inland refiner wished to exchange his ticket with a coastal refinery and, for simplicity, that the bartered oils were identical. In the absence of transactions cost, the minimum exchange ratio the inland refiner would accept would be 1 to 1; but the ticket allows him to buy oil that costs, say, $1.00 less per barrel than the domestic oil that the coastal refiner would otherwise buy. Consequently, if the price of domestic oil were $3.00 per barrel, competition among demanders of tickets would dictate an exchange ratio of one and one-third barrel of domestic to one barrel of foreign crude. The coastal refiner would then be indifferent between making this exchange or buying from his alternative supplier. However, if additional costs are associated with bartering, the ratio will never reach $1\,^1/_3$ to 1. For example, if the exchange rate thus becomes 1 to 1, the one-third barrel of oil, like the disappearing tobacco, must have been consumed in the process of exchanging.

To the extent that the buyer of a ticket incurs a positive cost of transacting, the exchange ratio will have to be somewhat less than $1\,^1/_3$, the amount of discrepancy reflecting that cost. Not all tickets are identical since transaction costs will differ among refiners. This means that a seller who subjects the buyer to a high transaction cost will have to accept less oil per ticket. So the differential in the amount of oil received per ticket reflects the differential in transaction costs; since buyers are not restricted in their choice of supplier, they all pay the same real price.

Competition implies that if transacting is costless, the value of a ticket must be identical among holders. Similarly, if positive transaction costs are proportional to the quantities transacted, if finding transactors is costless, and if ticket sellers are homogeneous, the value of a ticket will be identical between holders. Intramarginal buyers could exist, but the value of their marginal tickets would again be equated.

In the face of transactions costs, the buyer (who always has the option of purchasing domestic oil at price P_D) will never accept an exchange offer where the average cost of exchanging per ticket, CA, plus the sum of the pecuniary ticket price, t, and the foreign price P_f is greater than P_D . In fact, buyers will see to it that $P_D = t + CA + P_f$, where CA, and therefore t, are determined for each individual exchange. Since exchange is not mandatory, t

will never be negative; thus, if the cost of exchange becomes large enough, some tickets will go unused. The observation of unused tickets while others are exchanged is precisely the situation described earlier and would falsify the model which is based on costless transacting.

The prohibition on outright sale of tickets means that the pecuniary ticket price is zero, i.e., $t = 0$ (assuming no side payments). The exchange, however, is not constrained to a barrel-for-barrel swap. Consequently, the tickets' value may be restored by replacing the cash payment, t, with payment in kind. Thus, after correcting for quality differentials, the amount by which the physical exchange ratio exceeds unity gives the value of the exchange to the ticket seller. The record of actual oil exchanges provides a test of the traditional exchange model presented in figure 4. If we can discover points such as (Q^*, t^*), the traditional hypothesis is clearly falsified; and as might be inferred from the drama, there were such occasions.

The first major episode involving the nonuse of import rights occurred in 1967 when tickets for some 55 million barrels of crude oil went unused in Districts I–IV.[13] While these rights were not being exercised, others were exchanged. The phenomenon was again observed in 1971, prompting the *Oil and Gas Journal* to provide these insights:

> It's beginning to look like a significant batch of crude-oil import tickets may wind up in corporate wastebaskets at year-end.
>
> Mostly, they'll belong to small inland refiners who are finding traditional swaps with majors hard to come by. The exchanges that are made the rest of this year are going to be made at figures considerably below the long-time value of \$1.25–\$1.50/bbl. Recent swaps at 50¢/bbl now seem likely to remain the new high into the foreseeable future.

Later, the article adds:

> Total controlled imports of crude and finished products (excluding residual fuel oil in Dist. 1) for Dist. 1–4 were authorized at 1,450,000 b/d for 1971. But during the first half, crude and unfinished imports by refiners ran about 170,000 b/d below authorized levels from overseas sources.[14]

Let us return to figure 4 and ask if the counter-factual example (Q^*, t^*) can be reconciled with the hypothesis that generated (Q_O, t_O). If so, what does t^* measure? First, suppose that for some ticket holder the cost of exchanging, CA, is greater than t^*, that he owns $Q_O - Q^*$ tickets, and that for all others,

13. *Oil and Gas Journal,* January 8, 1968, p. 45. This figure represents approximately 13 percent of the 412 million barrels of crude imported during 1967.

14. Ibid., August 30, 1971.

their costs are less than t^*. Further suppose that the owner facing the next highest cost of exchange is just indifferent between accepting t^* and not exchanging. In this example, t^* (the fifty cent figure mentioned earlier) exactly measures the cost of transacting for the marginal exchanger who will lose all of the ticket's rental value to these transactions costs.

For t^* to be the measure of transaction cost, it is necessary that there be some tickets used while others are abandoned because it is not necessary that the costs of exercising the ticket equal its marginal value if the entire quota is being used. This might be the case, but it would only be coincidental that the transaction cost function intersected Q_O at t_O. For Q's smaller than Q_O, however, this marginal condition *must* be satisfied if some tickets are commanding a positive price.

To illustrate, if the entire quota were being used and the value of a ticket without transactions costs were $1.00, the exchange cost to the marginal seller could be anything from trivial to very close to $1.00. We can only infer from this observation that these costs are less than $1.00 in the most expensive transaction. That some rights remain unexploited implies that the prevailing ticket value is less than the costs of exercising the unused rights; thus, assuming a smooth cost function, the price of a ticket must exactly measure these costs to the marginal exchanger.

Given this type of transaction cost, we would expect to observe some effort to reduce it. Because oil companies owning refineries both inland and on the coast were free of such cost (they could use the quota earned inland in their coastal refineries) it would be attractive for such an organizational structure to increase in relative importance under the quota.

To test that hypothesis, District I–IV were divided into four areas—the Atlantic Coast (which refines most of the imports), the two main domestic production areas—the Gulf Coast and Inland, and the latter two combined.[15] Refining capacity was then classified according to whether it belonged to firms operating exclusively within one of the regions, the hypothesis being that, after the imposition of the quota, the share of interregional firms would expand. The data for the years 1958, 1965, and 1972 (summarized in table 6) show clearly that firms owning both coastal and inland refineries increased their share dramatically. Even as capacity was increasing in each of the regions, the absolute capacity held by firms operating in only one region actually declined. The change was both pervasive and large in absolute magnitude.

Our attempt to trace the progress of mergers in the industry was hampered by the paucity of available information, including the OIA's sometimes tardy

15. Even though the Gulf Coast is accessible to ocean tankers, since the price of oil in the Gulf Coast area is lower than in the Atlantic, Gulf Coast refiners will tend to use domestic oil. Consequently, it seems reasonable to combine them with inland refiners for present purposes.

TABLE 6

Share of Regional vs. Interregional Firms

	Atlantic Coast	Gulf Coast	Inland	Gulf Coast plus Inland
Total				
1958	1,345,300	2,786,850	3,917,962	6,704,812
1965	1,392,200	2,920,550	3,966,690	6,887,240
1972	1,414,600	5,191,955	4,201,364	9,393,319
Exclusive[a]				
1958	223,300	1,110,950[b]	2,207,762	4,226,212[c]
1965	140,000[d]	360,700	1,853,430	3,706,860
1972	140,000[d]	535,055	1,947,049	3,688,304
Nonexclusive[a]				
1958	1,122,000	1,675,900	1,710,200	2,478,600
1965	1,252,000	2,559,850	2,112,960	3,180,380
1972	1,274,600	4,656,900	2,254,315	5,705,015
Exclusive as percentage of total				
1958	16.60	39.86[b]	56.35	63.03[c]
1965	10.06[d]	12.35	46.72	53.82
1972	9.90[d]	10.31	46.34	39.27

[a] The capacity of a refiner with both Gulf Coast and Inland plants is included in the sum of the two but not in the individual catagories.
[b] Without Humble these figures become 828,950 and 29.75 percent.
[c] Without Humble these figures become 3,944,212 and 58.83 percent.
[d] All of this was capacity owned by Tidewater or Getty Oil.

determinations of ownership status.[16] Still, one is struck by the phenomenon that we were able to identify a few individual cases where ownership of inland refineries was acquired by coastal refiners. A list of these is presented in table

16. For the purposes of the oil import quota, *merger* was defined as the acquisition of a majority of the voting stock of a corporation by another corporation. When such a majority was held, the two companies were treated as one in the sense that their share of the quota was calculated on the basis of their total (combined) reported domestic crude processing. This definition was enforced by the Oil Import Administration in Skelly Oil Company's appeal, *Oil Import Digest,* Case No. Q-69, p. B-434. The Oil Import Appeals Board ruled that Getty Oil's ownership of 64 percent of the common stock of Mission Corp., which in turn owns 71 percent of Skelly, made Getty the owner of Skelly's tickets. This decision caused the essential revocation of Skelly's allotment because the size of Getty's historical allotment precluded them from also obtaining tickets based on refinery throughput.

7. However, more striking support of the hypothesis is found in the observation that we did not find a single counterexample in these merger data to the inland-coastal trend.

TABLE 7

Inland Refineries Owned by Refiners with Capacity on the East Coast, for Which Ownership Was Acquired after 1958

		Status in 1958		Status When Merged		
State	**City**	**Owner**	**Capacity**	**Year**	**Owner**	**Capacity**
Montana	Billings	Carter Oil Co.	24,000	1962	Humble	32,600[a]
Oklahoma	Duncan	Sunray	42,500	1971	Sunray	49,000
	Tulsa	Sunray	74,000	1971	Sunray	89,000
Utah	Salt Lake City	Salt Lake Refining Co.	41,000	1962	Standard California	43,000
	Salt Lake City	Utah Oil Refining Co.	30,000	1962	American	35,000
Total Capacity (1964)[b]						248,600 b/d

NOTE: In addition, 384,500 b/d were transferred from Standard Indiana to American Oil, and 75,000 b/d from Standard Texas to Standard California. At this writing, we are not certain whether these should also be included in the total since, even though they appear to be simple name changes, we do not know if this was the case for quota purposes.

[a] 43,000 by 1972.
[b] 1971 for Oklahoma transactions.

In 1970, imports from Canada were subjected to quota based on the historic import level during 1969. The distribution of that Canadian quota thus indicates the actual (or, more accurately, the stated) imports from Canada in 1969. As mentioned earlier, a refiner using Canadian rather than domestic oil did not earn import tickets, and due to the sliding scale, that penalty was proportionately more severe to smaller refiners. The results of our earlier tests, and the transaction-cost rationale accompanying them, suggest that the offer of extra tickets to smaller refiners was not a sufficient incentive to modify the size distribution of these refiners. The results for the Canadian data show the same pattern or, rather, lack of pattern as the earlier tests.

The implication of geographic integration yielded by the transaction-cost hypothesis also applies to these Canadian data. By the earlier reasoning we predict that American refiners owning coastal refining capacity will import less Canadian oil than those having inland capacity only. Figure 5 shows the fraction of Canadian oil in total oil input in 1969 for inland and integrated refiners. Refiners with no coastal capacity are mostly small; those with integrated capacity are mostly large. However, the ranges overlap, and it is clear that the latter imported less Canadian oil than the former. So the hypothesis is again confirmed: the barter requirement set by the quota imposed on refiners a real cost that favored those having joint inland and coastal capacity and thereby induced the expansion of that type of firm.

FIGURE 5

Midwest Refiners with Access to Canadian Crude

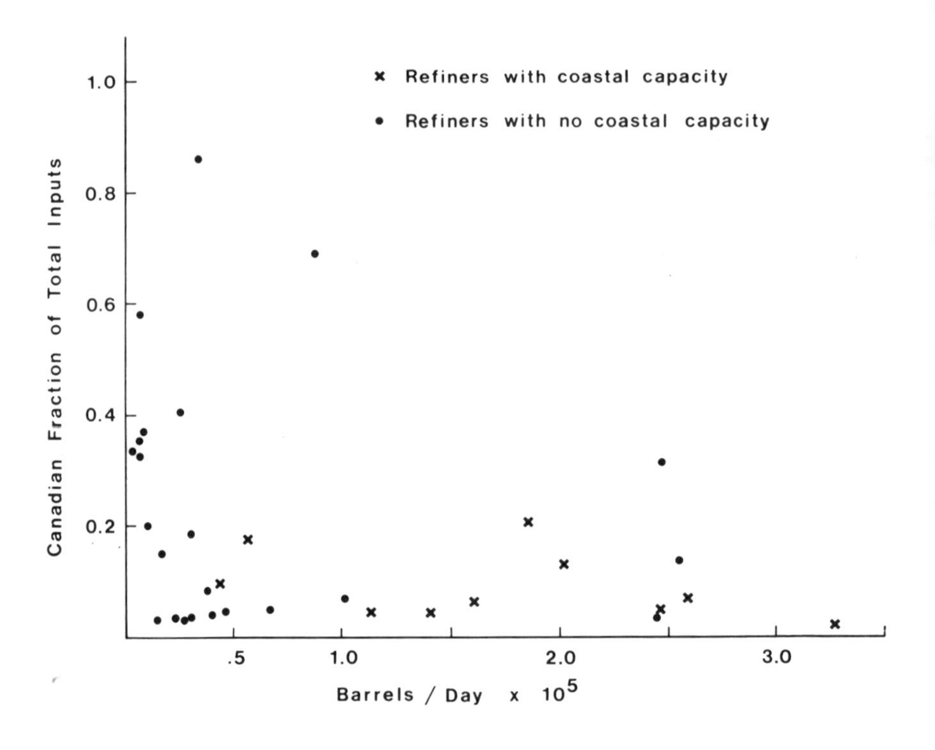

SOURCE: Data compiled from *Oil and Gas Journal* survey of refiners and various issues of *Oil and Gas Journal* oil allocation data.

6 • The Effect of the Quota on the Structure of Oil Prices

The central hypothesis that guided the initial phase of this research is that the quota would generate a shift in the relative prices of different grades of oil. With this framework and its implications, we set out to analyze the behavior of individuals constrained by the quota. Because prices are the heart of both theoretical and empirical research, this beginning appeared encouraging as well as somewhat novel in its use of prices. However, we soon learned that the data available to directly test the more obvious propositions were sadly lacking in quality. The deficicencies were so apparent that we now ask why these data were produced at all. At this writing no satisfatory explanation has been found.

In spite of this disappointment, we have utilized what price data are available in the formulation of a hypothesis regarding relative price change, so that some implications can be tested. Because of the inadequacy of these data the more interesting tests are found in the corollary hypotheses relating price changes to quantity adjustments reflecting the substitution of different types of oils affected by the quota. Taken together, the spectrum of oil grades constitutes the quality of domestic and imported oils.

What is designated as crude oil, whether domestically produced or imported, includes at any time a wide variety of crudes selling, on a per-barrel basis, over a substantial range of prices. Within each market, for any pair of oils the general equilibrium condition that must hold is that

$$MRS_{ij} = \frac{P_i}{P_j}$$

where MRS is the marginal rate of substitution, P is the per-barrel price, and i and j are two varieties of oil.

The equilibrium condition across markets must include the fact that the per-barrel cost of shipping oil is largely independent of its quality. Thus, in the absence of other constraints the prequota domestic prices for all traded oils had to exceed those of foreign prices by the same absolute amount. Given these two conditions, we begin by asking what can be said about the price structure subsequent to the quota's implementation.

To simplify, suppose for a moment that the relative shares of the high and low quality oils of domestic, foreign, and traded oils are the same. Suppose now that, due to the import quota, the volume—but not the composition—of traded oil has declined. The level of domestic prices must increase while foreign prices decline. Could domestic relative prices and foreign relative prices stay unchanged? Given the constant per-barrel shipping cost, the answer is: No, relative prices have to change. The proportionate increase in domestic prices implies a larger absolute per-barrel increase in the price of high priced oils, and conversely the fall in foreign prices implies a larger decline affecting high priced oil. The international price differential will thus become greater for high priced oils. In order to maximize the value of import rights, owners of these rights will be induced toward importing a higher proportion of high quality oils. At the point where the absolute price differences between domestic and foreign counterparts again are equalized, the substitution will cease.

If under the quota more than one quality of oil is imported, the postquota price structure has to show a narrowing of the gap between qualities inside the United States and a broadening of such a gap for foreign oils. If we use the subscripts d and f for domestic and foreign, respectively, and superscripts h and l for high and low quality crudes, then the ratio P^h_d / P^l_d will decline subsequent to the imposition of the quota; P^h_f / P^l_f will increase; and therefore $(P^h_d / P^l_d)/(P^h_f / P^l_f)$ will decline. It is easy to see that the propositions on these relative price changes are always satisfied, provided that each of the two crude oils in the pair under consideration is imported after the quota imposition.

The price data required to test the last proposition are difficult to find. Prices for crude oil as transacted in the market place are seldom published, and it is most difficult to construct consistent series for various crudes. The abundant list or posted prices notoriously fail to reflect a continuing variability in market prices. As it turns out, the evidence distilled from posted prices is totally inconclusive; whatever changes occur there provide neither confirmation nor refutation of the hypothesis.

A less direct but more fruitful approach is to formulate the hypothesis in terms, not of crude oil, but of oil products; in this area, the posted prices show more variability, and it may be hoped they conform to market changes. Specifically, and using basically the same reasoning as in the foregoing, we theorize that, since the light products are more valuable and more abundant in the high priced crude oils, the quota will cause the price of light products relative to that of residual oil to decline in the United States and to increase in foreign markets.

Figure 6 presents evidence where the U.S. price ratio is of marine diesel to residual at New York harbor, while the foreign ratio, at Curaçao (in the West Indies), is for gasoline relative to residual oil.[1] During 1958, the two price

1. The choice of oil products is dictated by availability of data and by their variability.

FIGURE 6

Ratios of Light Products to Fuel Oil

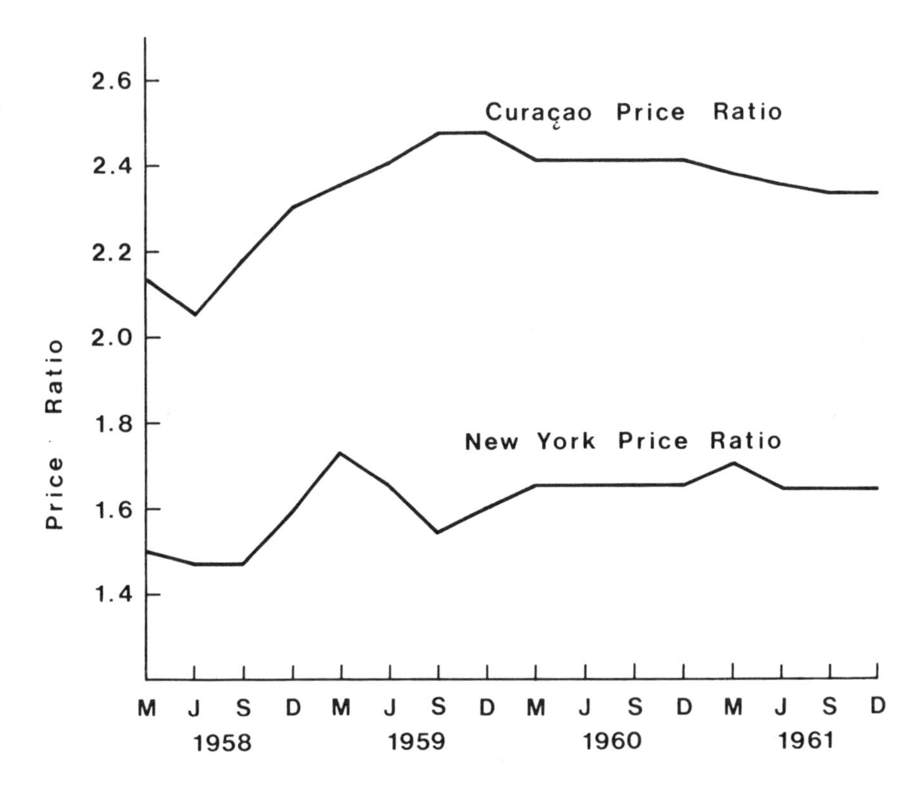

M = March
J = June
S = September
D = December

ratios moved more or less together. From March 1959, they diverged sharply, the relative price in the U.S. declining while in Curaçao it continued to increase. After September 1959, the two reversed their trends and continued to move in opposite directions to partly restore the pre-1959 ratios.

Noting that in general we would expect similar effects from forces that affect relative prices whether in the United States or in foreign markets, we feel it reasonable to attribute these observed divergences to the impact of the quota. Obviously the supporting evidence is not overwhelming. But given the difficulty of obtaining almost any market data for this industry, the results of the test are enlightening and encouraging.

A similar, but somewhat weaker, statement can be made with respect to quantities. The narrowing of the relative differential in domestic prices will provide a stronger incentive to expand output of lower quality oil. The actual

TABLE 8

Crude, Unfinished, and Products Imported into the United States

(thousands of barrels)

	Crude		Unfinished Oils	Products	
	Total	Canadian		Residual	Others
1950	173,920	5	7,512	123,037	4,084
1951[a]	177,356	473	5,265	121,969	3,390
1952	207,492	1,113	3,237	132,192	8,340
1953	237,908	2,650	2,922	134,824	7,142
1954	242,645	2,606	8,257	132,283	9,108
1955	294,096	16,395	6,616	155,458	13,832
1956	354,727	43,227	4,561	165,756	18,739
1957	385,802	53,804	1,959	176,021	27,324
1958	383,707	30,621	20,510	195,925	51,979
1959	384,597	33,902	23,127	223,414	43,029
1960	400,846	40,866	20,430	230,396	35,469
1961	411,968	65,819	25,802	240,106	39,255
1962	411,039	85,152	32,516	264,314	51,896[b]
1963	412,660	90,394	31,702	272,753	57,571
1964	438,643	101,607	32,587	295,771	59,698
1965[c]	452,040	107,762	33,706	345,187	69,299
1966	447,120	126,712	35,236	376,795	60,764
1967	411,649	150,409	35,225	395,939	95,361
1968	472,323	169,418	29,350	409,928	146,804
1969	514,114	203,298	38,766	461,611	168,071
1970	483,293	245,258	39,261	557,845	186,036
1971[d]	613,417	263,294	45,193	577,525	213,652

SOURCE: *Minerals Yearbook*(s).

[a] Asphalt listed as a separate product from 1951.
[b] Beginning in 1961, liquefied gases included explicitly.
[c] Beginning in 1965, small amounts of petrochemical feedstocks are listed in the *Yearbook*(s) but not reported here.
[d] Preliminary.

(relative) quantity will be larger only if the supply elasticity of low quality oil is not much less than that for high quality oil. In any case, data on crude production, classified by quality, are hard to come by.

Another set of data that conforms to this corollary hypothesis relates to the import of unfinished oils. This designation, like topped crude or unfinished gasoline, refers to partly processed crude oil from which usually the heavier (and lower valued) fractions have been removed at the field. The voluntary quota, which ignored oil products, failed also to cover such halfway products. We would expect, then, that the import of these high-valued oils would have increased as the result of an effective restriction on crude oil. That the voluntary quota was, indeed, effective along these lines is evidenced by an abrupt halt in the growth of crude imports in the late 1950s accompanied by continuing rapid growth in the importing of products, as seen in table 8.

FIGURE 7

Import of Unfinished Oils as Fraction of Crude

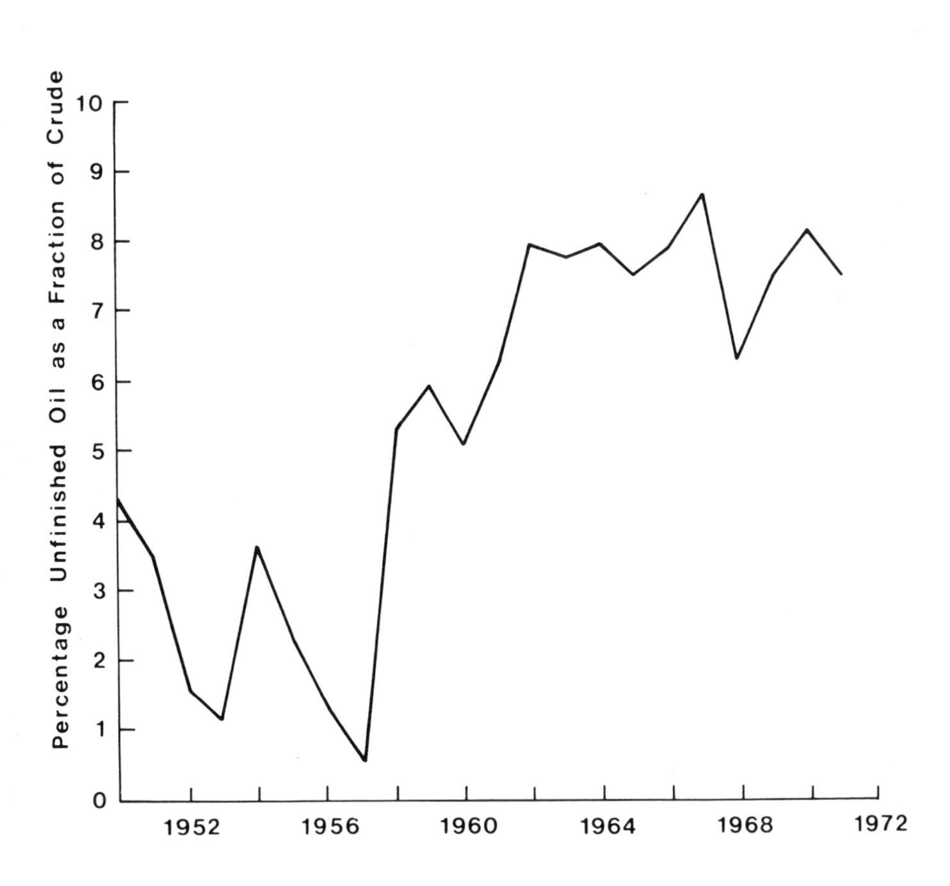

Figure 7 traces the time path of imports of unfinished oils as a fraction of crude oil. The jump in this percentage between 1957 and 1958 is most dramatic, and the new level was maintained in 1959. The value of unfinished oil was further demonstrated during the mandatory quota when the import of unfinished oils was restricted to a certain maximum percentage of crude imports and when, apparently, this maximum amount was always brought in.

The increase in this import level brings up another issue. By definition, unfinished oils require less processing than crude oil. Similarly, the higher the quality of crude, the less refining it will require to yield a given, and presumably optimal, set of oil products.[2] Our previous hypothesis would dictate that, subsequent to the quota, the imported crude oil would be of higher quality and would require less processing. We would expect, then, that one result of the quota would be the appearance of a lower ratio of labor in the refining of imported crude oil as against that needed for domestic crude.

Data are available by states on the refinery crude runs and their proportions of domestic crude and on the use of labor in refining.[3] In the following relation:

$$\left(\frac{L}{C}\right)_{1961} \Big/ \left(\frac{L}{C}\right)_{1958} = a + b\left(\frac{D}{C}\right)_{1961}$$

where L is man-years of labor, C is barrels of crude, and D is barrels of domestic crude, our hypothesis implies that $b > 0$. The form chosen for the dependent variable is designed to control for peculiarities within states. Two additional variables were considered. One, the average refinery size, was eventually dropped when it proved utterly insignificant. A second variable attempts to measure and account for variability in the quota-earning power of domestic crude, which is higher for smaller refiners. As the t ratio of this variable is close to one it has been retained, but its effect on the other variable is negligible.

The value obtained for b is 0.18 and its standard error is 0.078, giving a t ratio of 2.3. With twenty observations (17 degrees of freedom), this is a significant result. As the ratio of domestic to total crude increases from, say, 0.5 to 0.6, the regression results indicate that the 1961 ratio of labor to crude will be higher by 0.018 relative to the 1958 ratio.

---

2. For two crudes that differ only in the amount of impurities to be removed, the one with fewer impurities will be more expensive and will require fewer inputs of other factors in processing.

3. Data on labor are from *Annual Survey(s) of Manufacturers*, and from "Crude" in *Minerals Yearbook(s)*. In a few cases, for nondisclosure reasons, data are available by sets of states.

Consider some background issues. The mean number of men employed in refining in the twenty observations in 1961 was 6868; and the mean input of crude (in thousands of barrels) 143,501. In 1958, the number of men had been 7914 and the input of crude 138,240. Note, first, that the ratio of labor to crude declined by about 15 percent in the three-year span. This reflects a continuation of the trend in these series and is basically captured by the constant term in the regression of 0.74 (which is significantly less than one). In addition, note that the average number of barrels of all crude refined increased by only 4 percent. Presumably this is partly the effect of the quota. The quota, then, tended to reduce the size of the domestic refining industry not only in reducing the number of barrels of oil refined, but also by inducing the importation of oil requiring less refining.

7 • Summary and Conclusions

What economic effects are to be expected from an oil import quota? On conceptual grounds we argue that these would be not only widely varied but also quite different in nature from the results usually forecast in economic literature.

Because the term *crude oil* covers a whole range, any policy designed to regulate this resource must be complex but cannot be exhaustive. Market participants attempting to equate along any of the many margins open to them will give rise to a whole array of effects. The major effort in this study has been first to identify these margins, then to predict by economic reasoning the particular effects to be anticipated, and finally to test the hypotheses.

The scarcity of reliable quantitative data in this industry, particularly with respect to market prices, precludes any definitive measurement of the quota's total cost. Thus, of necessity, we confined our analysis to areas where other information could be summoned to test hypotheses. Where we could find little or no direct evidence we could only interpret behavior, utilizing the implied effects on resource allocation as viewed through the constraints of the quota.

Thus, for two reasons this study has included an elaborate and detailed section on the enforcement of the quota. First, even to define the import control program requires some understanding of its legal framework. Since that was never explicitly established, nor was its enforcement mechanism spelled out, one needs to search out the inherent de facto constraints it embodied. Second, the same information is needed in the attempt to define property rights to the rent generated by the quota—a basic ingredient in the search for adjustment costs and extensions of the hypothesis of relative price change. The observed relaxation of controls governing first residual, then distillate, and finally the lighter products was, we argue, to be expected because of the interaction between quota-induced changes in relative prices and the costs of establishing property rights to the quota rents.

Tests were conducted with respect to two major sets of activities: changes in the structure of the industry and changes in the volume and nature of refining.

The effect of the quota's sliding scale was more favorable to small refineries than to big ones. As expected, the size structure of the industry shifted toward smaller refiners on the West Coast (where virtually all refineries are near seaports). No such effect, however, is observed east of the Rockies.

Inland companies unable to use their quotas in their own refineries were permitted to barter them with coastal refiners. We argue, first, that this puts small refiners at a disadvantage and, second, that it lends a special advantage to a peculiar geographic integration between inland and coastal refineries.

Testing confirms the hypothesis of economies of scale in the barter arrangement and suggests that, on average, for each barrel bartered a substantial fraction of the rent generated by the quota restriction was dissipated by resource use necessary to effect the exchange.

The total dollar value of the dissipation was considerable. However, taking the reported values for import rights at face value in the periods when the entire allocation of import rights was exercised, little can be said of its magnitude, except of course, that these costs were less than $1.25 per barrel. There still must have been some cost of exchange during these periods, as the merger data confirm, and in the periods where a portion of the quota rights was abandoned, we do have a measure. Applying the fifty-cent figure to the some 170,000 barrels per day that were not imported during the two periods discussed, we arrive at a minimum of some $30,000,000 yearly lost because of the barter regulation. To this must be added the positive costs of exercising the tickets for the oil actually imported, including the costs of merger, and so on.

The merger hypothesis is a consequence of the cost of barter. In the absence of the quota we see no particular advantage for a refiner to own capacity both on the coast and inland, but because such a pattern of ownership obviated any need to use resources in bartering, it thereby achieved cost reduction. A test of the hypothesis does indeed reveal a marked shift toward coastal-inland integration.

Another predicted effect of the quota would be a change in relative prices both within crude oils and within oil products, reducing the relative spreads in the United States and increasing them abroad. The most direct test of the hypothesis, a comparison of crude oil prices, revealed nothing. The data available to us were posted prices; these are simply useless in studying price movements because they seldom reflect changing market conditions. We were somewhat more successful in testing the hypothesis by comparing product prices. Here there is evidence, as predicted, that the quota-induced price movements in the United States were in a direction opposite to those abroad.

Implicit in the hypothesis is the notion that, since the imported crude oil would now be of higher quality, less refining would be needed. A test reveals a lower ratio of labor per barrel of crude in states in which the ratio of imports was high.

The quota, then, reduced the overall level of refining not only through a decrease in number of barrels refined, but also by the importation of higher quality oil that required less processing. This "import of refining activity," it should be noted, was amplified by an increasing trend toward the importation of products—particularly light products—relative to crude oil.

In conventional analysis, a major (and, many argue, the *only*) impact of the quota on resource allocation is the creation of a welfare triangle. However, this clearly does not take into account the effects of the quota on refiners' size, structure, and the like. Moreover, the change in composition of imports toward higher quality crude means that the quota is less restrictive of quantity than is indicated by a mere count of barrels. Since the new pattern is not in itself efficient, a conceptually correct measure of welfare loss will give a smaller triangle than conventional analysis indicates, although to this must be added the resource cost of inefficiency in the mix of crudes, as well as the value of the resources expended in obtaining and enforcing quota rights. What would be the magnitudes of such discrepancies? Any reliable estimate would have to be supported by accurate price and quantity data. The case must rest on tested hypotheses.

Appendixes

The following appendixes contain more information than most readers may care to view. They are included in their lengthy form for two reasons. In addition to providing referenced material, we hope they will convey to the reader a feeling of the actual complexity of the quota by providing some detail of the regulations. Perhaps a glimpse of this complexity will serve to point out both some of the pitfalls and gains that the abstraction of theory requires.

It is hoped that, by viewing the sheer size of the definitional problem, the potential to adjust the regulations to either promote or penalize specific participants will be brought to bear more forcefully. Keep in mind that each segment of the regulation corresponds to some adjustment margin at which individuals differ in comparative advantage. These appendixes describe only a fraction of these margins.

Appendix A-1: Import Allocation Formulae

ALLOCATION OF IMPORT RIGHTS OF CRUDE AND UNFINISHED OILS (EXCEPT RESIDUAL FUEL OIL) TO REFINERS, DISTRICTS I–IV

Percent of Incremental Input to Be Imported in Various Periods

Period Input (thousand barrels/day)	1959 I	1959 II	1960 I	1960 II	1961 I	1961 II	1962 I	1962 II	1963 I
0–10	12	11.4*	13.0	11.6	11.5	11.1	11.6	12.0	12.5
10–20	11	10.4	11.6	10.8	10.5	10.2	10.7	10.2	10.7
20–30	10	9.5	10.8	9.8	9.6	9.3	9.7	10.2	10.7
30–60	9	8.5	9.7	8.8	8.6	8.3	8.7	8.2	8.6
60–100	8	7.6	8.7	7.9	7.7	7.4	7.7	8.2	8.6
100–150	7	6.6	7.5	6.9	6.7	6.5	6.7	5.2	5.3
150–200	6	5.7	6.5	5.9	5.7	5.5	5.7	5.2	5.3
200–300	5	4.7	5.3	4.9	4.7	4.5	4.7	5.2	5.3
over 300	4	3.8	4.3	3.9	3.8	3.7	3.8	5.2	5.3

Period Input (thousand barrels/day)	1963 II	1964 I	1964 II	1965 I	1965 II	1966	1967	1968 I	1968 II
0–10	12.5	14.0	15.0	17.0	18.0	18.0	20.0	19.0	18.8
10–30	11.5	11.9	11.2	11.6	11.9	11.4	11.4	10.2	10.1
30–100	9.2	9.3	8.9	9.2	9.4	8.9	8.0	6.7	6.6
over 100	5.4	5.45	5.28	5.53	5.64	5.26	4.28	2.74	2.7

Period Input (thousand barrels/day)	1969	1970	1971	1972	1973
0–10	19.5	19.5	20.0	21.7	21.7
10–30	11.0	11.0	12.0	13.0	13.0
30–100	7.0	7.0	7.0	7.6	7.6
over 100	3.0	3.0	3.5	3.8	3.8

NOTE: Prior to 1966 and in 1968, the allocation periods were six-month intervals—January 1 through June 30, and July 1 through December 31. These periods are indicated by I and II in this and the following tables.

*Estimated. The factor was not published in the *Federal Register* for this size class during this period.

ALLOCATION OF IMPORT RIGHTS OF CRUDE AND UNFINISHED OILS (EXCEPT RESIDUAL FUEL OIL) TO REFINERS, DISTRICT V

Percent of Incremental Input to Be Imported in Various Periods

Period Input (thousand barrels/day)	1959 I	1959 II	1960 I	1960 II	1961 I	1961 II	1962 I	1962 II
0–10	25	37.5	35.4	45.0	40.0	42.1	39.6	52.0
10–20	20	30.0	28.3	36.0	31.3	32.9	30.8	34.9
20–30	15	22.5	21.2	26.9	22.7	26.8	22.4	34.9
30–60	10	15.0	14.1	17.9	13.5	14.1	13.2	15.6
60–100	9	12.0	11.3	14.4	10.0	11.2	10.7	15.6
100–150	8	11.8	10.3	13.1	10.0	10.3	9.7	11.0
150–200	8	10.0	9.4	11.9	8.9	9.3	8.7	11.0
over 200	6	8.0	7.5	9.8	7.2	7.4	6.9	11.0

Period Input (thousand barrels/day)	1963 I	1963 II	1964 I	1964 II	1965 I	1965 II	1966
0–10	52.0	50.0	52.0	55.0	60.0	53.5	48.5
10–20	32.0	25.9	29.0	33.0	33.7	25.7	22.0
20–30	32.0	25.9	29.0	33.0	33.7	25.7	22.0
30–60	10.5	8.57	9.57	20.0	20.4	14.1	11.9
60–100	10.5	8.57	9.57	20.0	20.4	14.1	11.9
100–150	10.5	8.57	9.57	17.08	14.1	9.54	7.3
150–200	10.5	8.57	9.57	17.08	14.1	9.54	7.3
over 200	10.5	8.57	9.57	17.08	14.1	9.54	7.3

Period Input (thousand barrels/day)	1967	1968 I	1968 II	1969	1970	1971	1972	1973
0–10	8.5	45.0	44.3	40.0	40.0	40.0	60.0	67.5
10–30	18.2	11.0	10.7	9.3	9.3	9.3	15.0	16.9
30–100	9.8	5.2	5.0	4.3	4.3	4.3	5.0	5.6
over 100	6.0	2.2	2.2	1.9	1.9	1.9	5.0	5.6

NOTE: Eligibility—must have refining capacity in District V and have had such capacity for the period of one year ending three months prior to the allocation period for which the allocation is requested.

ALLOCATION OF IMPORT RIGHTS ACCORDING TO HISTORICAL CRITERIA

Based on import level under Voluntary Import Program.

Period	1959 I	1959 II	1960 I	1960 II	1961 I	1961 II	1962 I	1962 II	1963 I	1963 II
percent of import	80.0	80.0	80.0	80.0	66.5	70.3	65.9	70.0	67.5	57.0

Period	1964 I	1964 II	1965 I	1965 II	1966	1967	1968	1969	1970
percent of import	55.0	53.0	51.0	49.0	46.0	43.0	28.5	23.5	13.5

Based on overland imports during the Voluntary Import Program.

Period	1963 I	1963 II	1964 I	1964 II	1965 I	1965 II	1966	1967
percent of import	63.75	52.0	52.0	49.0	46.0	43.0	38.5	34.0

Period	1968[a]	1969[b]	1970[b]
percent of import	22.0	16.0	10.0

NOTES: These overland imports were not affected by the Voluntary Import Program; however, quotas are calculated as if these imports had been regulated.

As of 1/5/71—historical allocations revoked.

[a] 34 percent to be used if historical quota is less than input quota.
[b] Level not to fall below level based on input.

ALLOCATION OF UNFINISHED OILS

No allocation to be made that will entitle a person to import crude oils in excess of 10 percent of his total allocation.

OVERLAND IMPORTS

Overland imports were unregulated by the Voluntary Import Program, so that a person so importing was not counted under that program. On 12/15/62 an amendment was made to account for these allocations. If the import allocation based on input fell below a specified percentage of the oil that would have been imported under the Voluntary Import Program, the importer would then be granted an allocation based on his imports in the last period of the Voluntary Import Program.

PERCENTAGE OF OVERLAND IMPORTS, MADE DURING THE VOLUNTARY PROGRAM, THAT MAY BE IMPORTED UNDER THE MANDATORY PROGRAM

Period	**1963 I**	**1963 II**	**1964 I**	**1964 II**	**1965 I**	**1965 II**	**1966**	**1967**
Percent	63.75	57.0	55.75	52.0	49.75	46.75	42.25	37.75

Period	**1968**[a]	**1969**[b]	**1970**[b]
Percent	33.25	27.75	24.50

[a] If such an allocation is less than one calculated (for comparison) under the input quota (1), the person shall receive an historical allocation using the percentage rate 37.75.
[b] The historical allocation may not fall below the level of imports that would be allowed based on inputs.

FRACTION OF ALLOCATION UNDER VOLUNTARY IMPORT PROGRAM THAT MAY BE IMPORTED

Period	**1959 I**	**1959 II**	**1960 I**	**1960 II**	**1961 I**	**1961 II**	**1962 I**	**1962 II**	**1963 I**	**1963 II**
percent	80.0	75.7	75.7	75.7	72.0	70.0	72.8	70.0	67.5	65.0

Period	**1964 I**	**1964 II**	**1965 I**	**1965 II**	**1966**	**1967**	**1968 I**	**1968 II**	**1969**	**1970**
percent	63.0	61.0	59.0	57.0	54.0	51.0	45.0	45.0	40.0	30.0

NOTE: 1/5/71—Historical allocations revoked.

ALLOCATION OF IMPORT RIGHTS OF RESIDUAL FUEL OIL TO BE USED AS FUEL

3/11/61—Unless the historical quota is larger, allocation is to be based on terminal input(ti), using the year ending six months before the allocation period as a base period and using the following table:

Average ti (barrels/day)	Percent of ti as of 4/1/61	Percent of ti as of 4/1/62
0–1,000	100	100
1,000–5,000	41	45.1
5,000–10,000	29	31.8
10,000–30,000	16	17.6
over 30,000	10	11.0

3/19/63—Change the schedule to:

Average ti (barrels/day)	Percent of ti as of			
	4/1/63	4/1//64	4/1/65	4/1/66
0–5000	60.0	60.0	80.0	35.0
over 5000	26.7	27.25	40.0	15.0

As with the historical allocation, a separate schedule applies to persons without refinery capacity for the allocation periods beginning 4/1/63 and 4/1/64:

Average ti (barrels/day)	Percent of ti as of 4/1/63	Percent of ti as of 4/1/64
0–1000	100	100
1000–5000	70.0	70.0
over 5000	35.4	36.0

Appendix A-2: The Formal Definitions

This appendix was assembled by distilling the numerous Presidential Proclamations, Executive Orders, Bulletins, and Regulations which are found in the *Code of Federal Regulations, Title 3—The President,* and *Title 32A—National Defense Appendix.*

2-A. Person. Includes an individual, a corporation, firm or other business organization, or legal entity, and an agency of a state, territorial, or local government, but does not include a department, establishment, or agency of the United States.

2-B. Districts. Districts I–IV: The District of Columbia and all those States of the U.S. not in District V.

> **Amendment to 2-B:**
>
> **1/26/61**—Further defined as follows: (a) District I—the states of Maine, New Hampshire, Vermont, Massachusetts, Connecticut, Rhode Island, New York, New Jersey, Pennsylvania, Maryland, Delaware, West Virginia, Virginia, North Carolina, South Carolina, Georgia, Florida, District of Columbia; (b) Districts II–IV—those states in neither District I nor District V; (c) District V—the states of Arizona, Nevada, California, Oregon, Washington, Alaska, and Hawaii.

2-C. Crude Oil. "Crude petroleum as it is produced at the wellhead."

> **Amendments to 2-C:**
>
> **12/15/62**—Extended to include "liquids (under atmospheric conditions) that have been recovered from mixtures of hydrocarbons which existed in a vaporous phase in a reservoir and that are not natural gas products."
>
> **2/6/68**—Further extend the definition to include "initial liquid hydrocarbons produced from tar sands."

2-D. Finished products. Any one or more of the following petroleum oils, or a mixture or combination of such oils, which are to be used without further processing except blending by mechanical means.

(a) Liquefied gases. Hydrocarbon gases recovered from natural gas or produced from petroleum refining and kept under pressure to maintain a liquid state at ambient temperatures.

Amendments and revision to 2-D (a):
12/24/65—Adds specification of ethane, propane, and butane (but not methane) as types of liquid gases. Also specifies that to maintain the gases in a liquid state at ambient temperatures, the pressure level must exceed atmospheric pressure.
6/1/66—"Propane, ethane, butane, ethylene, propylene, and butylene (but not methane)" is the list of liquefied gases. Rest of definition is retained.
11/4/70—List maintained, with specification that liquefied gases be derived from crude oil or natural gas.
2/6/73—List maintained, with requirement only that liquefied gases be "derived by refining or other processing of natural gas, crude oil, or unfinished oils." Exception of methane is removed, but it is not listed as a liquefied gas.

(b) Gasoline. A refined petroleum distillate which, by its composition, is suitable for use as a carburant in internal combustion engines.

(c) Jet fuel. A refined petroleum distillate used to fuel jet propulsion engines.

(d) Naphtha. A refined petroleum distillate falling within a distillation range overlapping the higher gasoline and lower kerosenes.

(e) Fuel oil. A liquid or liquefiable petroleum product burned for lighting or for the generation of heat or power and derived directly or indirectly from crude oil such as kerosene, range oil, diesel fuel, topped crude oil, or residue.

(f) Lubricating oil. A refined petroleum distillate or specially treated petroleum residue used to lessen friction between surfaces.

(g) Residual fuel oil. A topped crude oil or viscous residuum that, as obtained in refining or after blending with other fuel oils, meets or is the equivalent of Military Specification MIL-F-859 for Navy Special Fuel Oil and any other more viscous fuel oil, such as No. 5 or Bunker C.

Amendments to 2-D(g):
8/5/67—A topped crude oil or viscous residuum which has a viscosity of not less than 45 seconds Saybolt universal at 100°F. minimum viscosity and which is to be used as fuel without further processing other than blending by mechanical means.
1/5/71—Requirement for topped crude oil or viscous residuum

unchanged; for crude is required only that it is to be used as fuel without further processing other than mechanical blending.

(h) Asphalt. A solid or semisolid cementitious material which gradually liquefies when heated, in which the predominating constituents are bitumens and which is obtained in refining crude oil.

(i) Natural gas products. Liquids (under atmospheric conditions) including natural gasoline, which are recovered by a process of absorption, adsorption, compression, refrigeration, cycling, or a combination of such processes, from mixtures of hydrocarbons that existed in a vaporous phase in a reservoir and which, when recovered and without processing in a refinery, otherwise fall within any of the definitions of products contained in Clauses 2 through 5 of this section D, "finished products."

2-E. Unfinished oils. One or more of the petroleum oils listed under "finished products," or a mixture or combination of such oils, which are to be further processed other than by blending by mechanical means.

2/F. Administrator. Administrator, Oil Import Administrator, Department of Interior, or a duly authorized representative. Later, "Director" means Director, Office of Oil and Gas, etc.

2-G. Importation, importing, import, imports, and **imported.** Include both entry for consumption and withdrawal from warehouse for consumption.

2-H. Refinery inputs. All crude oil, imported unfinished oils, natural gasoline mixed in crude oil, and plant and fuel condensaters mixed in crude oil which are further processed, other than by blending by mechanical means.

Amendments to 2-H:
6/23/62—List changed and simplified to: (i) crude oil; (ii) liquids (under atmospheric conditions) that have been recovered from mixtures of hydrocarbons which existed in a reservoir in a vaporous phase, with the exception of natural gasoline or plant condensate; (iii) imported unfinished oils which are further processed by other than mechanical means.
12/15/62—Introduction changed to: *Refinery inputs.* Refinery feedstocks . . . (listed) and the list is changed to: (i) crude oil; (ii) imported unfinished oils. See addition to "finished products" of "natural gas products."
12/15/66—"Crude oil" changed to "Crude petroleum as it is produced at the wellhead." Add two more categories: (iii) liquids which are recovered by a process other than absorption, adsorption, compression, refrigeration, cycling, or a combination of such processes from mixture of hydrocarbon that existed in a vaporous phase in a reservoir; (iv) ethane, propane, and butane which are recovered from natural gas and

which are chemically converted into other unfinished oils or finished products. (This category was added 6/1/66.)
6/7/68—Categories (iii) and (iv) dropped; "imported unfinished oils" changed to "unfinished oils imported pursuant to an allocation." Add (iii) unfinished natural gas products.
1/5/71—Introduction "refinery inputs" means feedstocks changed to "refinery capacity." Category (i) changed back to "crude oil."
12/22/71—Change "unfinished oils imported pursuant to an allocation" to: unfinished oils imported pursuant to an allocation if, and only if, (a) such imported unfinished oils are processed in a distillation unit with a resulting yield of at least two distinct finished products or unfinished oils, two of which must be equal to not less than 10 percent of the total charge of such imported unfinished oils to the distillation unit; or (b) such imported unfinished oils are subjected in catalytic or thermal conversion units to such processes as alkylation, coking, cracking, hydrofining, hydro desulfurization, isomerization, polymerization, or reforming; or (c) such imported oils are processed by dewaxing or deasphalting. Different grades or specifications of finished products or unfinished oils will not constitute distinct finished products or unfinished oils for [. . . provision (a)] of this subparagraph.

Exclusions: The following are specifically excluded from the definition of refinery inputs: (i) unfinished oils which have not been imported.

Amendments to exclusions in 2-H:
8/20/59—Add (ii) crude oil and unfinished oil imported into the United States by pipeline, rail, or other means of overland transportation from the country where they were produced, and which country, in the case of unfinished oils, is also the country of production of the crude oils from which the unfinished oils, is also the coutnry of production of the crude oils from which the unfinished oils were processed or manufactured.

6/22/60—The exclusion (ii) is reworded to specify that the exclusion refers only to counting such oil for computation of allocation under Section 10 or Section 11, rather than for purposes of eligibility.
6/23/62—Exclude (iii) natural gasoline or plant condensate.
6/23/62—Category (iii) is removed; see the change in the definition of crude oil (12/15/62). It seems that "Natural gasoline or plant condensate" is not part of crude oil.
12/15/62—Exclusion of (i) "Unfinished oils which have not been imported" is removed.
1/5/71—The Exclusion becomes "Crude oil or unfinished oils imported pursuant to paragraph (e), (f), or (h) of section 1A of Proclamation 3279, as amended, or with respect to refinery inputs in District V of

crude oil or unfinished oils imported pursuant to Clause (4) or Paragraph (a) of Section 1 of Proclamation 3279, as amended."

Section 1A: (e) allows importation of Canadian natural gas liquids without an allocation; (f) allows importation of ethane, propane, and butane produced in the western hemisphere without reducing quantities of crude oils and unfinished oils that may be imported; (h) allows limited importing of Mexican crude oil, unfinished oils and finished products.

Section 1(a) (4) is the earlier exclusion with regard to overland transportation.

2-I. Refinery capacity. Plant, which by further processing of crude oil or unfinished oils other than by blending by mechanical means, manufactures finished petroleum products.

Amendments to 2-I:

12/27/62—Changed to specify that to qualify as "refinery capacity" a plant must: (1) consist of stills, refining units, and equipment for separating or converting hydrocarbons, and storage tanks, pipeline, and pumps; (2) process crude oil or further process unfinished oils through the stills or units; and (3) manufacture two or more separate and distinct finished products, unfinished oils, or at least one finished product and one unfinished oil for a total yield equal to not less than 30 percent of total "refinery inputs."

6/7/68—Requirement changed to: (1) include equipment for separating or converting hydrocarbons to finished products or unfinished oils; (2) use crude oil as the predominant feedstock; and (3) convert for plant use as fuel in heating or generating power or for sale not less than 70 percent by weight of total refinery inputs into at least two separate and distinct finished products other than liquefied gases, each of which falls into a different one of the two categories specified in ["finished products"]—that is, gasoline, jet fuel, naphtha, fuel oil, lubricating oil, residual fuel oil, or asphalt—and each of which must be equal to not less than 4 percent by weight of total refinery inputs. Different grades or specifications of a finished product will not constitute separate and distinct finished products for the purpose of this definition.

2-J. Deep-water terminal (added 1/26/61, when special provisions were first made for residual fuel oil to be used as fuel). "An installation which (1) consists of bulk storage tanks, transfer, and handling of residual fuel oil; (2) is adjacent to waterways that permit the safe passage to the installation of a tanker rated at 15,000 cargo deadweight tons; and (3) has a berth that permits the delivery of residual fuel oil to be used as fuel into the installation by direct connection from a tanker rates at 15,000 cargo deadweight tons, drawing not less than 25 feet of water, and moored in the berth. Cargo deadweight tons represent the carrying capacity of a tanker, in tons of 2240 pounds, less the weight of fuel, water, stores, and other items necessary for use on a voyage."

Amendment to 2-J:
3/19/63—Specification (1) changed to further require minimum size of storage tanks as 100,000 barrels of operational capacity.

2-K. Petrochemical plant (added 6/1/66). A facility or a unit or group of units within a facility to which petrochemical plant inputs are charged and in which more than 50 percent (by weight) of such inputs are converted by chemical reactions into petrochemicals.

Amendments to 2-K:
12/31/66—Expands definition to allow for ". . . 50 percent (by weight) . . . or in which over 75 percent (by weight) of recovered product output consists of petrochemical."
6/7/68—Requires the manufacture, for plant use or sale, of one or more separate and distinct petrochemicals by chemical conversion of each separate petrochemical plant input feedstock stream which is claimed by an applicant as a basis for obtaining an allocation.

2-L. Petrochemical plant inputs (added 6/1/66). "Unfinished oils other than [exclusions]."

Amendment to 2-L:
6/7/68—"Feedstocks" changed to "a petrochemical plant" and including only: (i) crude oil; (ii) unfinished oils (except those unfinished oils specifically excluded) produced in Districts I IV and District V, and unfinished oils imported pursuant to an allocation.
Exclusions [as of 6-1-66]: (i) unfinished oils which are imported into the United States by pipeline, rail or other means of overland transportation from the country where they are produced, which country is also the country of production of the crude oil from which the unfinished oils were processed or manufactured; (ii) unfinished oils produced in a petrochemical plant as a byproduct in the manufacture of petrochemicals and subsequently recharged to the same chemical plant in which they were produced.

Amendments to exclusions in 2-L:
6/7/68—Wording of exclusion (ii) tightened to exclude interplant shipment of produced petrochemicals within a firm; (iii) added to preclude transaction—sales, purchases or changes—designed to avoid exclusion (ii).
9/11/69—Add (iv) benzene or toluene or any xylene derived from crude oil, liquefied gases, or natural gas products which meets the distillation specifications of the ASTM standard specification for that chemical but

which subsequently has been recycled and mixed with other hydrocarbons, commingled, or purposely debased.

2-M. Petrochemicals. Organic compounds or chemical elements other than unfinished oils or finished products, produced from petrochemical plant inputs by chemical reactions.

Amendment to 2-M:
6/7/68—carbon or organic compound (other than finished product or unfinished oils) produced from . . . (as above).

2-N. Petroleum oils. As used in "finished products" and "finished oils," liquid hydrocarbons derived from crude oil.

Appendix B: Some Posted Prices

Standard Oil of California Prices

Gravity	Kettleman Hills Coalinga*	Buena Vista, Elk Hills (Shallow), Midway Sunset	Coles Levee, Elk Hills (Stevens Zone)	Long Beach (Signal Hill), Huntington Beach*
10–10.9	....	$2.09	....	$2.10
11–11.9	$2.16	2.15	...	2.15
12–12.9	2.22	2.21	. .	2.20
13–13.9	2.27	2.27	....	2.25
14–14.9	2.33	2.32	....	2.31
15–15.9	2.40	2.39	....	2.38
16–16.9	2.46	2.46	....	2.45
17–17.9	2.53	2.53	....	2.51
18–18.9	2.59	2.60	....	2.58
19–19.9	2.66	2.68	....	2.65
20–20.9	2.72	2.75	....	2.72
21–21.9	2.79	2.82	....	2.78
22–22.9	2.85	2.89	$2.64	2.85
23–23.9	2.91	2.96	2.71	2.91
24–24.9	2.98	3.03	2.78	2.98
25–25.9	3.04	3.10	2.85	3.05
26–26.9	3.10	3.17	2.92	3.12
27–27.9	3.16	3.24	2.99	3.19
28–28.9	3.21	3.31	3.05	3.26
29–29.9	3.26	3.37	3.11	3.32
30–30.9	3.31	3.43	3.17	3.38
31–31.9	3.36	3.48	3.23	3.45
32–32.9	3.41	3.53	3.29	3.51
33–33.9	3.46	3.57	3.35	3.57
34–34.9	3.51	3.61	3.41	3.62
35–35.9	3.56	3.65	3.47	3.67
36–36.9	3.61	3.69	3.52	3.72
37–37.9	3.66	3.73	3.57	3.76
38–38.9	3.71	3.77	3.62	....
39–39.9	3.76	3.81	3.67	....
40–40.9	3.81	3.85	3.72	....

NOTES: Sample price postings from Platt's Oilgram Price Service, February 21, 1973, Volume 51, No. 36-B.

All gravities above those quoted take highest price offered for field specified.

Richfield, Wilmington	El Segundo, Torrance	Inglewood	Pyramid Hills Cymric*	Kern River Kern Front*
. . . .	. . . .	$2.02	. . . .	$2.00
. . . .	. . . .	$2.10	$1.97	2.05
$2.19	. . . .	2.18	2.07	2.10
2.26	. . . .	2.26	2.17	2.15
2.33	$2.33	2.33	2.27	2.22
2.41	2.40	2.41	2.34	2.29
2.47	2.46	2.48	2.41	2.36
2.54	2.53	2.55	2.48	2.43
2.61	2.59	2.63	2.55	2.49
2.68	2.66	2.70	2.63	2.56
2.75	2.72	2.77	2.70	2.63
2.82	2.78	2.84	2.78	2.69
2.89	2.84	2.91	2.85	2.76
2.96	2.90	2.98	2.93	2.82
3.03	2.96	3.05	3.00	2.89
3.10	3.02	3.12	3.07	2.95
3.17	3.08	3.18	3.15	3.02
3.23	3.14	3.23	3.23	3.09
3.26	3.20	3.28	3.28	3.15
3.33	. . . .	3.33	3.33	3.21
3.38	. . . .	3.37	3.37	3.27
3.43	. . . .	3.40	3.42	3.33
. . . .	. . . .	3.44	3.47	3.39
. . . .	. . . .	3.48	3.52	3.44
. . . .	. . . .	3.52	3.57	3.49
. . . .	. . . .	3.56	3.61	3.54
. . . .	. . . .	3.60	3.65	3.59
. . . .	. . . .	3.64	3.69	3.64
. . . .	. . . .	3.68	3.73	3.69
. . . .	. . . .	3.72	3.77	3.74
. . . .	. . . .	3.76	3.82	3.79

*Price postings also listed by Mobil Oil Corp. In some instances Mobil listings include additional prices for gravities above and/or below those shown.

Union Oil Company of California Prices

Gravity	Kettleman Hills Jacalitos Coalinga	Midway-Sunset Buena Vista*	Long Beach (Signal Hill), Huntington Beach*
10–10.9	. . . .	$2.09	. . . .
11–11.9	. . . .	2.15	. . . .
12–12.9	$2.22	2.21	$2.20
13–13.9	2.27	2.27	2.25
14–14.9	2.33	2.33	2.31
15–15.9	2.40	2.39	2.38
16–16.9	2.46	2.46	2.45
17–17.9	2.53	2.53	2.51
18–18.9	2.59	2.60	2.58
19–19.9	2.66	2.68	2.65
20–20.9	2.72	2.75	2.72
21–21.9	2.79	2.82	2.78
22–22.9	2.85	2.89	2.85
23–23.9	2.91	2.96	2.91
24–24.9	2.98	3.03	2.98
25–25.9	3.04	3.10	3.05
26–26.9	3.10	3.17	3.12
27–27.9	3.16	3.24	3.19
28–28.9	3.21	3.31	3.26
29–29.9	3.26	3.37	3.32
30–30.9	3.31	3.43	3.38
31–31.9	3.36	3.48	3.45
32–32.9	3.41	3.53	. . . .
33–33.9	3.46	3.57	. . . .
34–34.9	3.51	3.61	. . . .
35–35.9	3.56	3.65	. . . .
36–36.9	3.61	3.69	. . . .
37–37.9	3.66	3.73	. . . .
38–38.9	3.71	3.77	. . . .
39–39.9	3.76	3.81	. . . .
40 and above	3.81	3.85	. . . .

NOTES: Sample price postings from Platt's Oilgram Price Service, February 21, 1973, Volume 51, No. 36-B.

Gravities in excess of those specified take highest price quoted for that field.

Richfield Wilmington*	Torrance	Inglewood	Edison Kern Bluff Kern Front	Cuyama South
....	....	....	$2.00	....
....	....	....	2.05	....
$2.19	....	....	2.10	....
2.26	....	....	2.15	....
2.33	....	....	2.22	....
2.41	$2.40	....	2.29	....
2.47	2.46	....	2.36	....
2.54	2.53	$2.55	2.43	....
2.61	2.59	2.63	2.49	....
2.68	2.66	2.70	2.56	....
2.75	2.72	2.77	2.63	....
2.82	2.78	2.84	2.69	....
2.89	2.84	2.91	2.76	....
2.96	2.90	2.98	2.82	$2.71
3.03	2.96	3.05	2.86	2.78
3.10	3.02	3.12	3.95	3.85
3.17	3.08	3.18	3.02	2.92
3.23	3.14	3.23	3.09	2.99
3.28	3.20	3.28	3.15	3.05
3.33	....	3.33	3.21	3.11
3.38	....	3.37	3.27	3.17
3.43	....	3.40	3.33	3.23
....	....	3.44	3.39	3.29
....	....	....	3.44	3.35
....	....	....	3.49	3.41
....	....	....	3.54	3.47
....	....	....	3.59	3.52
....	....	....	3.64	3.57
....	....	....	3.69	3.62
....	....	....	3.74	3.67
....	....	....	3.79	3.72

*Price postings also listed by Mobil Oil Corp. In some instances Mobil listings include additional prices for gravities above and/or below those shown.

Bibliography

Adelman, M. A. *The World Petroleum Market.* Baltimore, Md.: Johns Hopkins University Press, 1972.

Annual Survey of Manufacturers. U.S. Bureau of the Census. Washington, DC.: Government Printing Office.

Barzel, Yoram. "An Alternative Approach to the Analysis of Taxation." Journal of Political Economy 84 (December 1976).

———. "A Theory of Rationing by Waiting." Journal of Law and Economics 17 (April 1974).

Bhagwati, J. "On the Equivalence of Tariffs and Quotas," in his *Trade, Tariffs, and Growth.* Cambridge, Mass.: MIT Press, 1969.

Borrows, J. C., and Domencich, T. A. *An Analysis of the U.S. Oil Import Quota.* A Charles River Associates Research Study. Lexington, Mass.: Heath Lexington, 1970.

Cheung, Steven N. S. "A Theory of Price Control." Journal of Law and Economics 17 (April 1974).

Coase, Ronald H. "The Problem of Social Cost." *Journal of Law and Economics* (October 1960).

Dam, Kenneth W. "Implementation of Import Quotas: The Case of Oil." *Journal of Law and Economics* 20 (April 1971).

Kirby, Aubrey M., Jr. "The Chemical Industry and the Oil Import Program." *U.S. Petrochemicals.* Edited by A. M. Brownstein. Tulsa, Okla.: Petroleum Publishing Co., 1972.

Kochin, Levis. "Monopoly Profits and Social Losses." Unpublished manuscript, University of Washington.

Krueger, A. O. "The Political Economy of the Rent Seeking Society," *American Economic Review* 63 (3) (June 1974).

Oil and Gas Journal.

Oil Import Digest. Washington, D.C.: National Petroleum Refiners Association.

Petroleum Economist. London. (Formerly: Petroleum Press Service.)

Shaffer, Edward H. *The Oil Import Program of the United States.* New York: Praeger, 1968.

Stern, R. M. "Tariffs and Other Measures of Trade Control: A Survey of Recent Developments." *Journal of Economic Literature* 11 (September 1973): 857–88.

Tullock, Gordon. "The Welfare Cost of Tariffs, Monopolies and Theft." *Western Economic Journal* (June 1967), no. 3.

U.S. Bureau of Mines. *Minerals Yearbook*. Washington, D.C.: Government Printing Office.

U.S. Cabinet Task Force on Oil Import Control. *The Oil Import Question: A Report on the Relationship of Oil Imports to National Security*. Washington, D.C. (1970).

U.S. General Services Administration, National Archives and Records Service, Office of the Federal Register. *Code of Federal Regulations: National Defense, Appendix, 32A,* Chapters X–XI. Washington, D.C.: Government Printing Office. (Numerous volumes.)

———. *Code of Federal Regulations: Title 3—The President*. Washington, D.C.: Government Printing Office. (Numerous volumes.)

Index

Adelman, M.A., 5n
Allocation of import rights, *see* Crude oil; Fuel oil
Arthur D. Little, Inc., 48
Asphalt, 42-43, 83
Atlantic Richfield, 35

Baker and Bolts, 36
Barzel, Yoram, 26n, 27n, 29n
Bhagwati, J., 1n
Bonded fuel, 8-9
Borrows, J. C., 1n

Canada, 7-8, 10-14, 52n, 63-64
Carson, Captain, 17, 19-20
Cheung, Steven N.S., 27n, 29n
Coase, Ronald H., 30n
Committee to Support the Mandatory Oil Import Program, 36
Commonwealth Oil, 43
Consumers Union, 36
Crude oil: allocation of, 7, 13, 51-64; allocation sliding scale, 76-80; heterogeneity of, 2-3, 72; posted prices, 5, 65-67, 88-92; price variation, 2-3, 39, 65-67

Dam, Kenneth W., 7, 13n, 46n, 49n
Delta Refining Co., 17
Domencich, T. A., 1n
Dow Chemicals, *see* Petrochemicals

Eisenhower, Dwight D., 8

Free Trade Zones, *see* Petrochemicals
Fuel oil, 8, 37-39, 41, 82; allocation sliding scale, 80

Gasoline, 43, 82
Getty Oil, 62n
Gulf Oil, 43

Hess, Leon (Hess Oil and Chemical Corp.), 40 42, 50
Historical import base, 7, 78-79; dissipation involving, 51-53
Hoehn, Elmer L., 36
Hudson, John J., 42, 42n, 43n

Ikard Bill, 19
Independent Producers Association of America (IPAA), 19, 32, 46, 50
Industrial Solvents Corp., 31n

Johnson, Lyndon B., 8, 36-37, 42, 50
Jones Act, 50n

Kennedy, John F., 8
Kennedy, Robert, 41-42
Kirby, Aubrey M. Jr., 17n, 48n
Kochin, Levis, 27n
Kreuger, A. O., 27n

Long, Russell B., 41

Mandatory oil quota: definition of, 1-2, 81-87; effects on refining, 2, 70-71, 73; inception, 4; national defense, 4, 12, 22; national defense districts, 7, 8, 14-15, 76-77, 81; potential rents, 23-30, 34, 51, 57-64, 72
Merger, 53-56, 61-64
Mexico, 7, 14

Nation Petroleum Refiners Association (NPRA), 46

New England governors, 36, 46
Nixon, Richard M., 8, 46
Northeast Petroleum Corp., 41
Northville Dock Corp., 38

Occidental Petroleum Corp., *see* Petrochemicals
Office of Defense Mobilization, 16
Office of Emergency Preparedness, 42
Oil and Gas Journal, 2, 41, 42, 50, 60
Oil Import Administration (OIA), 5, 31, 33, 36
Oil Import Appeals Board (OIAB), 5, 8, 31, 33, 37, 40, 42, 49, 61-62
Oil products, 7, 17, 69-70, 81
Overland exemption, 7
Overland imports, 7, 10-15, 79. *See also* Canada; Mexico

Petrochemicals: Atlantic Richfield, 35; BTX, 47, 48n, 86; Dow Chemicals, 45; feedstocks, 86; Free Trade Zones, 9, 44-46; Occidental Petroleum Corp., 46; Phillips Petroleum, 49; Standard Oil Co. (N.J.), 49; Standard Oil of Indiana Petrochemical, 47-48; Sun Oil, 49; Union Carbide, 9 10, 43n, 44, 49
Petroleum Economist, 2-3
Phillips Petroleum, *see* Petrochemicals
Proxmire, Senator, 48
Puerto Rico, 8, 45, 49

Refining, *see* Mandatory quota
Refining allocation, *see* Crude oil

Shaffer, Edward H., 14n, 43n
Skelly Oil Company, 62n
Sliding scale, *see* Crude oil; Fuel oil
Special Committee to Investigate Crude Oil Imports, 16
Standard Oil Co. (Indiana), 36
Stanford Research Institute, 48
Stern, R. M., 1n
Sun Oil, *see* Petrochemicals
Superior Oil, 17

Texas-American Asphalt Co., 32, 36
Texas Independent Producers and Royalty Owners Association (TIPRO), 32, 46
Tickets, 3, 20, 27-28, 32-33, 42n, 48, 53-61, 63-64
Transaction costs, 4, 53-64, 73
Tullock, Gordon, 27n

Udall, Interior Secretary, 37-38, 40-43, 48, 49
Unfinished oils, 7
Union Carbide, *see* Petrochemicals
U.S. Cabinet Task Force on Oil Import Control, 52n

Venezuela, 17
Virgin Islands, 8, 41, 50
Voluntary oil quota, 7-8, 16-21, 79; base for historical allocation, *see* Historical import base